Engaging Indonesia

Critical Dialogues on Culture and Society

Engaging Indonesia brings together current debates in social and cultural studies relating to the study of Indonesia, with the aim of giving voice to both leading and emerging Indonesianists. Critical and interdisciplinary, the books in the series contribute to knowledge formation by addressing pressing issues facing the country - cultural, social, and political - from contemporary and historical perspectives. Works may include research-based monographs, edited volumes, handbooks, and translations of both recent, and seminal/historical works. The series is particularly welcoming of topics relating to issues of urban and environmental change, to questions of identity and place, to problems of governance and conflicts, to the politics of representation, language, and performances, and to the global and postcolonial contexts in which these are addressed. Special consideration is given to studies that cover underrepresented regions and marginalized voices, as well as those that seek to recover lost archives, uncounted political events, or forgotten historical figures. Relevant to students and scholars from a broad range of disciplines in the humanities and social sciences with a particular interest in researching Indonesia, the series champions intellectual engagement with the archipelago's rapidly changing landscape - from within, and beyond.

Abdul Halim Iskandar

The Methods of Measuring the National Sustainable Development Goals in Indonesia

SDGs Desa (Volume 2)

Buku OBOR

Springer

Abdul Halim Iskandar
Jakarta, Indonesia

ISSN 2948-2909 ISSN 2948-2917 (electronic)
Engaging Indonesia
ISBN 978-981-96-0926-0 ISBN 978-981-96-0927-7 (eBook)
https://doi.org/10.1007/978-981-96-0927-7

Jointly published with Yayasan Pustaka Obor Indonesia
ISBN of the Co-Publisher's edition: 978-623-321-181-9

The original submitted manuscript has been translated into English. The translation was done using artificial intelligence. A subsequent revision was performed by a third party to further refine the work and to ensure that the translation is appropriate concerning content and scientific correctness. It may, however, read stylistically different from a conventional translation.

Translation from the Indonesian language edition: "SDGs Desa: Metodologi dan Pengukuran (Trilogi SDGs Desa #2)" by Abdul Halim Iskandar, © Yayasan Pustaka Obor Indonesia 2024. Published by Yayasan Pustaka Obor Indonesia. All Rights Reserved.

This Springer imprint is published by the registered company Springer Nature Singapore Pte Ltd.
The registered company address is: 152 Beach Road, #21-01/04 Gateway East, Singapore 189721, Singapore

Preface

Ex-President Joko Widodo received aspirations from the community that village funds would be increasingly beneficial for all villagers, especially for the poor. Conceptually, the benefits of development are measured through the outcome, adding it to the input measure. The latest human development studies, which are just as determined to equalise development for all, without leaving any citizens behind, are the Sustainable Development Goals (SDGs) from 2016 to 2030.

The tasks of the state and society as stipulated in the articles of the 1945 Indonesian Constitution are similar to the goals and targets of the SDGs, such as, equitable access to education and healthcare facilities, democracy and freedom of speech, addressing poverty, and the opportunity to practise religion in daily life.

The Village Law also mandates that development must be able to reduce poverty, provide infrastructure, improve the quality of life, and preserve the environment. These goals are also detailed in the SDGs.

Specifically, the aim of village development in the 2005–2025 National Long-Term Development Plan is to equalise development, so that it ultimately supports economic growth. Absolute development equalisation is certainly in the form of 'no one left behind', an adage used to simplify the determination of the SDGs. Indonesia is one of the initiators of the SDGs at the UN, and then applied it in Indonesia, as decided in Presidential Regulation Number 59/2017 on the acceleration of Sustainable Development Goals, which was reinforced with Presidential Regulation Number 111/2022.

Considering the mandate of the constitution, the mandate of legislation, the mandate of the President, and the ratification of the SDGs, I have decided to apply them at the village level, localising the SDGs into SDGs Desa. If not localised, the Sustainable Development Goals document shows that the role of the village is limited only to SDGs goal 1, to eradicate poverty, and SDGs goal 10, which is to improve self-reliant villages in order to reduce regional disparities. Only two SDGs originally considered the role of the village. However, in reality, 91% of the smallest government areas are villages, and 71% of the Indonesian population lives in villages.

Localising the SDGs at the village level expands the role of the village to all SDGs. This is truly in line with the proportion of the large role of villages in Indonesia.

The localisation of SDGs into SDGs Desa aims to clarify the direction of village development, facilitate its implementation in the field, and facilitate the measurement of results, benefits, and impacts of development.

The SDGs icon is localised in the SDGs Desa so that villagers immediately understand it. For example, the goal of reducing poverty is more easily understood by the savings icon, while the global icon is often misunderstood as similar to the family planning programme.

Localisation of the statement formulation is done so that the ultimate goal to be achieved by the village is easily understood. For example, the statement 'Village Without Poverty' is easier to understand than 'Ending all forms of poverty everywhere'.

The SDGs Desa are placed in the cultural context of villages in Indonesia. For this reason, SDGs Desa goal 18 has been specifically added: 'Dynamic Village Institutions and Adaptive Village Culture'. This additional goal shows the uniqueness of village development in Indonesia and ensures that cultural identity and the rights of traditional communities are respected, in line with the development of the times and civilisation. The village promotes culture by guaranteeing the freedom of the community in maintaining and developing its cultural values. Likewise, the village maintains the local language as a part of the cultural wealth of the nation.

In their entirety, SDGs Desa have 18 goals, with 222 indicators, which cover aspects of citizenship, village territory, and village institutions. These become the direction of village development policy until 2030. The stages of village development start from data collection, planning, implementation, and include monitoring and evaluation. Specifically, this regulation ensures the sovereignty of the village over its own data.

The democratisation of SDGs Desa data is shown first by 'data from the village', that is, information is collected from the village government, Neighbourhood Units (Rukun Tetangga—RT), families, and individual village residents. The adage 'data by the village' is realised by the formation of the Village Data Collection Volunteer Working Group, which are data collectors from residents in the same RT. This participatory census ensures data validity, as it is hard to lie to a data collector who is also a neighbour. The adage 'data for the village' is realised by placing only the data guardian in the village as the sole data holder for By-Name-By-Address (BNBA) residents and families, as well as RT and village environment information.

Information technology is not only used to collect data but also to process that data to produce information related to village activity recommendations. Additionally, information technology assists in monitoring the results of SDG implementation at the village level. Currently, through the website www.sid.kemendesa.go.id, village residents know the exact number of poor and extremely poor residents, the number of residents who need health and employment insurance, and so on. Village residents also get information on recommended activities that each village needs. Specifically, the data holders in the village can review up to BNBA residents and families of potential beneficiaries of activities.

Although the algorithm successfully processes, analyses, and concludes village development recommendations, the final decision still lies with the village deliberation forum. All algorithms only act like experts for the village.

The SDGs can be interpreted as the highest achievement of the concept of human development since the formulation of the development concept in 1945 until now. The same applies to its localisation in SDGs Desa. However, as per scientific norms so far, readers still have room to continue innovating in the realm of SDGs Desa. We all still face the challenge of formulating the best policies to achieve the expected development targets by 2030, namely to fulfil the needs of life for all people (leaving no one behind).

Jakarta, Indonesia
September 2023

Abdul Halim Iskandar

Contents

About the Author

Abdul Halim Iskandar is a former Minister of Villages, Development of Disadvantaged Regions, and Transmigration (2019–2024). He grew up in a large family and completed his primary and secondary education at Mambaul Ma'arif Islamic Boarding School in Jombang, East Java. He Then he completed his undergraduate education in the fields of philosophy and sociology of education at Yogyakarta State Institute of Teacher Training and Education (now Yogyakarta State University), and his master's degree in educational management at Malang State Institute of Teacher Training and Education (now Malang State University) in 1995.

Village life is nothing new for him. As well as being born and growing up in a village, since graduating right until receiving the mandate as chairman of the Local People's Representative Council of East Java Province, most of his life has been given to serve NU (Awakening of Islamic Theologian Organization), Islamic boarding schools, and villages.

As a villager, young Halim firmly believed that change must be driven by young people.

Therefore, as a member of the legislature for two periods from 1999 to 2009 and as chairman of Local People's Representative Council of Jombang Regency, as well as two periods from 2009 to 2019 as leader of Local People's Representative Council of East Java Province, he fights for important issues such as the development of villages, education, health, the Islamic boarding school, and micro, small and medium enterprises (MSMEs). A policy in the form of regional regulations regarding the empowerment of village communities in both Jombang and East Java Province was born when he was leader of the Local People's Rrepresentative Council, long before the Village Law was born. This led to the award an honorary doctorate in community empowerment management by Yogyakarta State University in 2020.

In Yogyakarta, he was active in the Indonesian Islamic Student Movement (PMII); in NU he was active as a branch manager at NU Regional Administrator of East Java (2018–2023); and in National Awakening Party (Partai Kebangkitan Bangsa/PKB), he is the chair of PKB Regional Administrator of East Java Province (2011–2027) and is one of the chairs of PKB National Administrator (2019–2024).

In educational institutions, his career began when he returned from studying in Yogyakarta. From 1987 until now he has been administrator of the Denanyar Jombang Islamic Boarding School; from 1987 to 1989 he served as a teacher at MAN (Islamic Senior High School) Mambaul Ma'arif, Jombang; and he was lecturer at the Hasyim Asy'ary Islamic Institute (Ikaha, now Hasyim Asy'ari University/Unhasy) Tebuireng from 1987 to 2004. He was appointed Dean of the Tarbiyah Faculty from 1993 to 1997, initiated the birth of the Sultan Agung Jombang Vocational School, and became the Head of the Sultan Agung Vocational School, from 1996 to 2000. Recently, he was Chair of the Advisory Board of the State University of Surabaya (2018–2022) and General Chair of the Advisory Board of Yogyakarta State University (2020–2023).

He carried out his work in the health sector with Kyai and NU activists in Jombang. He is also one of the founders of the NU Hospital Jombang and is currently commissioner of NU Hospital in Jombang (from 2019 until now).

To communicate his ideas and thoughts, he has written books and many articles in various print and electronic media. In 2012 with Thoriqul Haq (now Regent Head of Lumajang) he wrote *New Jihad Resolution: NU's Ijtihad Gives Birth to PKB*. In 2019, he published a collection of his articles titled *Negeri Holopis Kuntul Baris*. His articles have been published in *Kompas*, *The*

Jakarta Post, Jawa Pos, Media Indonesia, Republika, Kontan, Kedaulatan Rakyat, among others.

Chapter 1
Prologue: Position of Methodology and Measurement

In Indonesia, the village occupies two strategic positions. First, the governmental area comprising 75,265 villages is dominant, covering 91% of local administration in Indonesia; the remaining 9% is the urban governmental area. Certainly, as local administration, the village government directly interacts with residents every day. Similarly, for the residents themselves, the village government is the everyday representation of the state. Moreover, the proportion of village government is truly dominant in Indonesia.

The second strategic position is the very large and dominant village population. The Ministry of Home Affairs reports that 71% of residents holding identity cards live in villages, while 29% are urban residents. The Village Development Index (IDM) 2023 census collected a figure of 214 million village residents.

Because of these two strategic positions, solving village problems, innovating for acceleration of progress, and other positive actions for the village, are essentially building Indonesia. Therefore, it is appropriate that ex-President Joko Widodo was determined to build Indonesia from the periphery and from the village.

1.1 SDGs Desa

The first book in the trilogy, titled *Accelerating the Achievement of the National Sustainable Goals in Indonesia: SDGs Desa (Volume 1)* (Iskandar 2020), explained the background and the new concepts of the Village Sustainable Development Goals (SDGs Desa). The book provided SDGs Desa as the main narrative, becoming a discourse discussed in various forums and media, and directing village development since the second semester of 2020.

Crucially, SDGs Desa show the direction of village development policy and community empowerment. The formulation of this policy direction is based on the Village Information System. Eighteen goals make up SDGs Desa, namely:

A. H. Iskandar, *The Methods of Measuring the National Sustainable Development Goals in Indonesia*, Engaging Indonesia, https://doi.org/10.1007/978-981-96-0927-7_1

(1) Village without poverty
(2) Village without hunger
(3) Healthy and prosperous village
(4) Quality village education
(5) Involvement of village women
(6) Village with clean water and sanitation
(7) Clean and renewable energy village
(8) Equitable village economic growth
(9) Needs-based village infrastructure and innovation
(10) Village without gaps
(11) Safe and comfortable village residential areas
(12) Environmentally conscious village consumption and production
(13) Climate change responsive village
(14) Village caring for the marine environment
(15) Village caring for the land environment
(16) Village of peace and justice
(17) Partnerships for village development
(18) Dynamic village institutions and adaptive village culture.

In practice, SDGs Desa can be divided into two core values, namely citizenship value and territoriality value. Citizenship and territoriality analysis is very useful for sharpening the pattern of village development, such as the orientation of human development towards the goals within the scope of citizenship, and the development of physical facilities in the village and rural areas within the scope of territoriality.

The SDGs Desa numbers 1–6 are classified as citizenship values, so can be used to carry out community empowerment activities. The measurement includes information on each community member, By-Name-By-Address (BNBA), so that no village residents are left behind in development. Because there are universal resident and family indicators, the total measurement results from 214 million village residents can be directly read as a contribution to individual and family data at the national and global level.

Territoriality value is included in the SDGs Desa numbers 7–18, which refers to the environment or context of the village where village residents carry out their daily activities. Territoriality value becomes the basis for village development. To better align with the needs of village residents and the selection of village development locations, territorial data is collected at the Neighbourhood Units (or *Rukun Tetangga*—RT) level. Adaptation of indicators is needed to better match the conditions of 827,464 RTs and also 75,265 villages in Indonesia. For example, toll roads are adapted to village roads, and indicators of airfields and railway tracks are eliminated.

For comparison, SDGs at the global and national levels (Bappenas 2020a, b, c, d) are usually divided into four pillars of development, namely the pillar of social development, the pillar of economic development, the pillar of environmental development, and the pillar of legal and governance development. Analysis of these four SDGs pillars prioritises aspects of human development, while the analysis of the two

Table 1.1 Locus of SDGs Desa goals in analysis 2: core values and four pillars

Core values	Pillars			
	Social development	Economic development	Environmental development	Legal development and governance
Citizenship	1, 2, 3, 4, 5	–	6	–
Territoriality	–	7, 8, 9, 10, 17	11, 12, 13, 14, 15	16, 18

core values of SDGs Desa refers to the strength of the method. Both can be used together to strengthen the method of holistic human development analysis (Table 1.1).

1.2 Not Ideal, but Concrete

If we only delve into the first book of the SDGs Desa Trilogy, understanding could merely converge on abstract thought. Moreover, the concept of SDGs Desa is a grand idea that encompasses almost all aspects of village life. Grand ideas are typically considered difficult to implement in the field.

However, the reality is just the opposite. SDGs Desa are not an abstract idea, but a concrete one to be implemented in the field. This is shown by the ease of field data collection. Then, based on that data, recommendations for village development and community empowerment are issued. The implementation of these recommendations takes the form of development activities, which involve villagers, thus proving that SDGs Desa are a practical concept. Looking at the stages of thought-data collection-implementation, the relevance of this second book, *SDGs Desa: Methodology and Measurement*, is apparent. First, it shows that SDGs Desa are a practical concept, not an ideal concept that is impossible to implement in the field. Their practicality is shown by the collection of data in the field: it can be measured from the field.

Second, the measurement of SDGs Desa can be carried out by the villagers themselves. If the methodology of SDGs Desa were too difficult, it would certainly require high expertise, which cannot be carried out by ordinary villagers. However, the methodology of SDGs Desa is deliberately carried out in a participatory manner: from, by, and for the village. That is, it is based on the data needs of the village that have already emerged, data collection is carried out by the villagers themselves, and development and empowerment recommendations are directed towards each village.

Third, the data collected is microdata, which is most effective and efficient when compiled by the villagers themselves. The method is participatory census, not conventional survey. To obtain citizenship data, data updating is carried out by conducting a census of all families and villagers. As for obtaining territorial data, data updating takes the form of a census to all RTs. As a result, the data to be collected will include the conditions of the villagers and their families by name and by address, as well as the diversity of the village's territorial conditions at the location closest to the residents' settlements. Such data enhances the effectiveness of empowerment activities development or building in areas directly inhabited by residents.

Fourth, for the first time, all data analysis and recommendations are aimed at specific villages. In this way, each individual village benefits. Of course, the recap data can still be compiled for analysis and recommendations for village development at the sub-district, district, provincial, and national levels. Recap data like this is common, but it doesn't actually provide benefits for the village itself. What the village needs is data at the level of RTs, families, and residents. Now, with the ability to compile analysis and recommendations for each village, the entire SDGs Desa data from 75,265 villages across the archipelago, can be directly useful for each village. It can be stated that this is the largest village study every year.

Weber (1978) used the ideal type as a qualitative method. It is composed using reasoning or ratio, by continuing the field conditions to the most extreme point that might be thought of. For example, the ideal type of poverty refers to the absence of income, assets, food, housing, and social networks. This cannot be found in the field, because once found then the poor person must have died. However, the ideal type of poverty allows people to imagine degrees of poverty and determine the causes that vary from village to village. The ultimate goal is to facilitate poverty alleviation programmes.

However, in everyday life, ideal refers to things outside of life, which cannot be found in life. Ideal is even used sarcastically to label concepts that cannot be implemented.

To counter the sarcasm of the ideal, a concept must be found in the field. To be usable, the field findings must be distinguished in terms of their highs and lows, measured by the scale between them. With measurements like this, the position of something in relation to something else is then known. This then allows residents to increase or decrease their position.

For example, consider SDGs Desa number 1: Village without Poverty. In the field this can be found by calculating family assets, such as roof, walls, floor, doors and windows of the house, ownership of valuable items that can be exchanged for money, as well as family expenditure and income. SDGs Desa number 1 can also be shown by the percentage of children from poor families who attend primary school, secondary school and university. Also, poverty is shown by the use of health facilities available in the local village.

By showing the operational definition in the field, which can be seen and then recorded, the SDGs Desa are not a floating concept but have concrete foundation of field data. As soon as the field data is analysed, the characteristics and causes of poverty in a village can be known, which can differ from other villages.

The results of this analysis then become the basis for implementing activities to reduce village poverty. At this point, the SDGs Desa also show their practicality, that is, they are used to formulate activities in the field, in order to overcome shortcomings or enhance the potential of the village.

1.3 Microdata for Use

National surveys produce macrodata, where the analysis is comprehensive at the national, provincial, and district levels. So far, macro analysis has been the basis for formulating policies, programmes and village development activities at the national, provincial, and district levels.

The main obstacle arises during implementation, as what is actually needed is microdata. The characteristics of microdata are that it is collected from the field, so the names and addresses of anyone whose data is collected can be traced, usually in the form of a census of all residents or the entire region.

To compare the two, Indonesia has macro poverty data issued by BPS twice a year, in March and September. Meanwhile, the current micro poverty data is in the form of Integrated Social Welfare Data (*Data Terpadu Kesejahteraan Sosial*—DTKS), which is released by the Ministry of Social Affairs every January and July.

The macro poverty data from BPS (Central Bureau of Statistics) in September 2020 concluded that there are 15.51 million rural residents below the poverty line (BPS 2021a, b, c, d). However, this sample figure is not a cumulative total of each poor person in the village. It is not possible to find the poor people in the village from the data used to conclude this figure.

In reality, it is also not easy to find and determine each poor person in the field, when the basis of calculation is still determined from the centre. The slow updating of field data also contributes to the low accuracy. The Integrated Database (the name used before it was changed to DTKS) only captures 23% of the poor. The rest are errors in including the non-poor group or not including the truly poor group.

Recognising this weakness, the SDGs Desa data should ideally be microdata. This refers to data that can be referred to in the field (By-Name-By-Address—BNBA). Information is collected from each village resident, family, RTs, and village data from the field. Each summary of supra-village data, from district data, city/county data, provincial data, to national data, can be traced back to the field in remote villages.

1.4 Annual National Village Study

The SDGs Desa data is participatory; it is collected from the village itself, carried out by the village itself, and its main use is for the village itself. The village statistical data that has been collected as far as the national level has been used to formulate village development at the national, provincial, and city/county levels. The analysis of this data forms the basis for formulating medium-term development plans, annual development plans, and programmes and activities.

However, ironically, this village-level data cannot be used by the village itself in formulating the Village Medium-Term Development Plan (RPJM Desa), the Village Government Work Plan (RKP Desa), and the formulation of activities in the Village Revenue and Expenditure Budget (APB Desa). This is because the village data collects cumulative information at the village level, whereas the village needs data from each level: RTs, families, and individuals. The selection of road construction locations, for example, requires data on the quality of roads in each RT, then choosing the RT with the condition of dirt roads or asphalt roads that have been damaged. The construction of Uninhabitable Houses (RTLH) requires a list of families living in slum houses. Scholarships for the poor require data on school-age children who are not in school.

The collection of detailed data from within the village since the 1970s is known as village studies (Lipton and Moore 1980). So far, it has taken the form of in-depth research in one or several villages. Data is collected from village residents, families in the village or small community areas within the village. With this detailed data, analysis can be compiled within the scope of the village. A theory about a village has the potential to be developed, thus facilitating the decision-making process to develop the village.

So far, village studies have been considered expensive, time-consuming, and not very useful for the country. Therefore, village studies have only been conducted by academics in certain locations. Such methods make it difficult for village studies to form the basis of regional and national village development policies, and ultimately do not help many villages within a country.

In fact, it is possible to conduct national village studies, even on an annual continuous basis. Throughout 2020 it has been proven that villages can gather data from 8.1 million poor families within a period of two months. The method is, the data collector compiles a list of poor families in each neighbourhood, which usually number from three families to dozens of families. Then all this data is collected up to one village. The data is validated through village deliberation forum, to determine these poor families.

The success of the census strategy run for each neighbourhood gives hope for conducting national village studies. By using an application on a mobile phone, the most difficult data collection effort is at the beginning of the data collection. The data that has been collected remains listed in the application, so that in the next data collection, the only effort is in updating the existing data. This greatly facilitates the village.

References

Bappenas. 2020a. *Metadata indicator of economic development pillar*. Jakarta: Bappenas.

Bappenas. 2020b. *metadata indicator of law and governance development pillar*. Jakarta: Bappenas.

Bappenas. 2020c. *Metadata indicator of environmental development pillar*. Jakarta: Bappenas.

Bappenas. 2020d. *Metadata indicator of social development pillar*. Jakarta: Bappenas.

BPS. 2021a. *Poverty and inequality*. [internet]. Available at: https://www.bps.go.id/subject/23/poverty-and-inequality.html.

BPS. 2021b. *Percentage of poor population (P0) by province and area*. [internet]. Available at: https://www.bps.go.id/indicator/indicator/view_data/0000/data/192/sdgs_10/1.

BPS. 2021c. *Percentage of poor population September 2020 increased to 10.19 percent*. [internet]. Available at: https://www.bps.go.id/pressrelease/2021/02/15/1851/persentase-penduduk-miskin-september-2020-naik-menjadi-10-19-persen.html#:~:text=Jumlah%20penduduk%20miskin%20pada%20September,88%20persen%20pada%20September%202020.

BPS. 2021d. Poverty profile in Indonesia. In *Official statistics news*, vol. 24, no. 16/02.

Iskandar, and A. Halim. 2025. *Accelerating the Achievement of the National Sustainable Goals in Indonesia: SDGs Desa (Volume 1)*. Berlin: Springer.

Lipton, M., and M. Moore. 1980. *Methodology of rural studies in developing countries*. Jakarta: Social Sciences Foundation.

Weber, M. 1978. *Economy and society* . California: University of California Press.

Chapter 2
Philosophical Framework of SDGs Desa

Law Number 6/2014 affirms the close relationship between the constitution and the village. The 1945 Constitution, Article 18B, Paragraph 2 ensures the state recognises customary law units in accordance with the principles of the Unitary State of the Republic of Indonesia (NKRI). The state also respects traditional rights that exist according to societal developments.

The constitution creates a strong foundation for villages to carry out governance and development towards a fair and prosperous society. The Village Law also contains the objective that the village should play a role in realising the ideals of independence based on the 1945 Constitution.

The noble aspirations of village development are traced normatively through the principles of recognition and subsidiarity. On the ground, they are pursued in the achievements of the 18 SDGs Desa targets.

2.1 Constitution, Recognition, Subsidiarity

The best classic, and richly descriptive writings about the village were originally composed by Mpu Prapanca in the fourteenth century (Prapanca 2017). Interestingly, the title is *Negara Kertagama*, but it is also known as *Desa Warnana.* The meaning of *Negara Kertagama* as a statehood that receives cultural, religious, and sacred tradition support, is manifested in the field in the diversity of villages also known as Desa Warnana. The constitution and the village seem to have both been inscribed by Mpu Prapanca.

The villages listed in *Negara Kertagama*, have been traced on a journey by writer and historian, Hadi Sidomulyo. Over six centuries, the inhabitants of the villages visited by Mpu Prapanca remain dynamic, because they are supported by traditions and cultures that are not eroded by time.

A. H. Iskandar, *The Methods of Measuring the National Sustainable Development Goals in Indonesia*, Engaging Indonesia, https://doi.org/10.1007/978-981-96-0927-7_2

The principle that the state must protect and empower the village is crystallised by recognition in the Village Law. The original rights of the village, which manifest the physical existence of the village, are recognised by the state. This is demonstrated by the fact that all villages that were formed before 2014 have a unique area code number.

The state's respect for traditional rights is summarised in the principle of subsidiarity. In the context of regional autonomy, the village has officially recognised local authority, so it is able to protect the interests of village residents. The village can also determine its own authority, guided by the rules of authority arranged by the regional government.

The norms to realise these noble ideals of independence are materially supported through village funds (Dana Desa—DD). The regional government also channels village fund allocations (Alokasi Dana Desa—ADD), financial aid, and tax revenue sharing. More than Rp400 trillion in village funds flow from the state general cash account to the village cash account. The aspiration to help village residents prosper is marked by a decrease in the number of severely disadvantaged and disadvantaged villages.

In 2015, there were still 33,592 severely disadvantaged villages and 13,453 disadvantaged villages. However, by 2023, the number of seriously underdeveloped villages had significantly reduced to 4,850 villages, and the number of underdeveloped villages had decreased to 7,154.

This means that 35,041 villages or 84% have been improved. Clearly, villages play a role in moving towards the aspiration of the constitution to advance public welfare and enhance the life of the nation.

2.2 Constitution and SDGs Desa

The details of Indonesia's aspirations in nation-building are recorded in the articles of the 1945 Constitution. On this point, the SDGs Desa facilitate a practical, methodological, measurable path, so that the village has a plan of action and stages to realise the mandate of the constitutional articles.

Since the end of 2015, the United Nations (UN) has crystallised all human experiences in nation-building into the Sustainable Development Goals (SDGs). As one of the initiating countries, Indonesia immediately implemented it in the form of Sustainable Development Goals (TPB) in Presidential Regulation 59/2017.

Its localisation at the village level is known as the SDGs Desa. The SDGs Desa are a measurable policy direction to realise the 1945 Constitution and the Village Law. President Joko Widodo specifically directed the government to ensure that all village residents benefit from village funds and village development. This is immediately reminiscent of the SDGs principle: 'No One Left Behind'. There should not be a single person overlooked, there should not be a single village resident who cannot enjoy the results of village development.

Therefore, SDGs Desa are used as an integrated effort to accelerate the achievement of sustainable development goals. The SDGs Desa contain 18 goals. The SDGs Desa goal 1: Village without Poverty, and goal 2: Village without Hunger, as well as goal 3: Healthy and Prosperous Village, aim to ensure the achievement of Article 34 of the 1945 Constitution. The details are to care for poor and neglected children, to include them in the social security system, and to empower the vulnerable and disabled community. In addition, the village must provide adequate healthcare facilities and public service facilities.

The SDGs Desa goal 16: Village of Peace and Justice, includes targets that correspond with Articles 27, 28A, 28D, 28G, 28H, and 28I of the 1945 Constitution, namely the equality of village residents before the law, while receiving legal protection, equality, and justice, without discrimination. This also includes protection of personal property rights, free from torture and treatment that degrades human dignity.

The SDGs Desa goal 1: Village without Poverty, goal 8: Equitable Village Economic Growth, goal 9: Needs-based village infrastructure and innovation, and the goal 10: Village without Gaps are opening the path for villages to realise Articles 27, and 28D of the 1945 Constitution. Village residents get the opportunity for decent employment and livelihood. This is indicated by fair and decent rewards and treatment in employment relations. The village economy itself is directed towards economic democracy with the principles of togetherness, justice efficiency, sustainability, environmental insight, independence, and by maintaining balance national economic progress and unity.

Sustainable village development requires institutional strength, as outlined in SDGs Desa goal 16: Village of Peace and Justice, goal 17: Partnerships for Village Development, and goal 18: Dynamic Village Institutions and Adaptive Village Culture. This ensures the implementation of the constitution Articles 28, 28B, and 28C, that is, the freedom to associate, gather, and express opinions verbally or in writing. Every citizen also has the right to advance themselves in collectively fighting for their rights to build the village.

Indeed, SDGs Desa goal 16: Village of Peace and Justice; and goal 18: Dynamic Village Institutions and Adaptive Village Culture are to ensure the implementation of the constitution Article 28D, that is, every citizen has the right to obtain equal opportunities in village governance.

This is supported by another target in SDGs Desa goal 16, to implement the constitution Article 28F. That is, every citizen has the right to communicate and obtain information from the village to develop their personal and social environment, and has the right to search, obtain, possess, store, process, and convey information in the village.

Families and children in the village are protected according to SDGs Desa goal 1: Village without Poverty, goal 2: Village without Hunger, goal 3: Healthy and Prosperous Village, goal 4: Quality Village Education, goal 5: Involvement of Village Women, and goal 16: Village of Peace and Justice. The purpose of these SDGs Desa is the gateway to realise Article 28B, so that everyone has the right to form a family and continue it with their offspring. Meanwhile, every child has the right to survival, growth, and development as well as the right to protection from violence and discrimination.

SDGs Desa goal 4: Quality Village Education, goal 16: Village of Peace and Justice; and goal 17: Partnerships for Village Development, have given a boost to realising the constitution Articles 28C, 28E, and 31. The purpose of the constitution is that village citizens receive education and benefit from science and technology, arts and culture, embrace religion and worship according to their religion.

SDGs Desa goal 18: Dynamic Village Institutions and Adaptive Village Culture, ensures the implementation of the 1945 Constitution Articles 28E and 29, that is every village citizen has the right to embrace religion and worship according to their religion. Village citizens also have the right to freedom of belief, and to express thoughts and attitudes, in accordance with their conscience.

Settlement and environmental issues are covered in SDGs Desa goal 3: Healthy and Prosperous Village, goal 11: Safe and Comfortable Village Residential Areas, goal 12: Environmentally Conscious Village Consumption and Production, goal 13: Climate Change Responsive Village, goal 14: Village Caring for the Marine Environment, and goal 15: Village Caring for the Land Environment. The above SDGs can pave the way for the achievement of the constitution Article 28H, so that village citizens live prosperously in body and soul, have a decent place to live, and a good and healthy living environment, as well as being able to obtain health services.

Finally, village culture is preserved through SDGs Desa goal 16: Village of Peace and Justice and goal 18: Dynamic Village Institutions and Adaptive Village Culture. This implements the 1945 Constitution Articles 28I and 32, that is cultural identity and the rights of traditional communities are respected in line with the development of the times and civilization. The village promotes culture by guaranteeing the freedom of the community in maintaining and developing its cultural values. Similarly, the village maintains the local language. as a cultural wealth of the nation.

Achieving the goals of the SDGs Desa essentially realises the objectives of the Indonesian nation's constitution. This is the philosophical foundation of the SDGs Desa.

Reference

Prapanca, M. 2017. *Desawarnana: Adaptation of Kakawin Nagarakertagama.* Jakarta: Komunitas Bambu

Chapter 3
Methodology

To strengthen the position of the measurement of SDGs Desa, the theoretical foundation needs to be profound, which usually suggests the highest abstraction of a philosophical position. In 1954 J.F. Ferrier made a distinction between ontology as the basis of what exists, with epistemology as the path to truth (Hunnex 2004). Epistemology is the theory of knowledge that asks how humans can know.

The truth theories used in epistemology include the theory of correspondence, coherence, and pragmatism (Agusta 2014). In the theory of correspondence, an idea or proposition is true if it accurately and sufficiently resembles or represents reality, or if there is at least a correlation between the two. The coherence theory explains that an idea or proposition is only true when it is consistent with, or required by, the totality of truth, in which the idea or proposition becomes part of it. In the pragmatic theory, an idea or proposition is true when it functions, satisfies, or is capable of doing something.

In the past, these epistemological divisions could be separated and were not usually not united (there were mutually exclusive). Survey methods, for example, were considered incompatible with qualitative methods. Survey methods fall under the correspondence epistemology, while many qualitative methods operate within the coherence epistemology.

However, a mixed method that combines survey methods with qualitative methods (Creswell and Clark 2018) is now recognised. The aim is to understand reality more broadly and more deeply at the same time.

3.1 Epistemology and Methodology of SDGs Desa

Although it also uses a mixed method, the measurement of SDGs Desa moves between the correspondence and pragmatic epistemologies. The use of census methods refers to the correspondence epistemology. The pragmatic epistemology

A. H. Iskandar, *The Methods of Measuring the National Sustainable Development Goals in Indonesia*, Engaging Indonesia, https://doi.org/10.1007/978-981-96-0927-7_3

is demonstrated by the participation of the villagers themselves in conducting the census of their neighbours, who are known to be close, thus increasing the validity of the data. The data also fully belongs to each village itself, not the central government, as the highest authority.

Methodology differs from methods and techniques (Agusta 2014). Technique indicates data collection instruments, data processing, and data analysis, for example, questionnaires, in-depth interviews, participant observation, and document analysis. Method is a field approach with a set of these techniques, such as survey methods, census, and case studies. In a census, for example, the main instrument is a questionnaire, while data verification can be done with an interview.

Methodology is a level above methods and techniques. Methodology becomes a metatheory for methods and techniques. Methodology places a series of methods and techniques according to the equivalence of assumptions about the reality being excavated, the process of information gathering, and the actions carried out by the data collector.

The SDGs Desa methodology is derived from epistemology to achieve the truth of the SDGs Desa. Specifically, the truth of the SDGs Desa is achieved when:

(1) The targets of the 18 SDGs Desa goals are achieved. This shows the teleology in the SDGs Desa to achieve the set goals.
(2) The mechanism of cooperation between parties in achieving these targets is fulfilled. The participatory methodology is read from the agreement of various parties to cooperate in achieving the targets that are covered in all the goals of the SDGs Desa
(3) Truly realised when applied at the village level. Evidence in the field becomes correspondence that shows the concept of SDGs Desa is truly achieved in the field.

3.2 Methodological Consequences of Achieving SDGs Desa Goals

The achievement of truth through the achievement of the targets of the 18 SDGs Desa has consequences for methodology, methods, and data collection techniques. The characteristic of achieving SDGs Desa is formulated as achieving all positive indicators and eliminating all negative indicators. For example, SDGs Desa 3: Healthy and Prosperous Village, for citizenship dimensions it achieves positive indicators if 100% of citizens have health insurance, while the prevalence of (Tuberculosis) TB decreases to 0%. As for the territorial dimension, it would include the elimination of malaria in 100% of RTs.

The formulation of SDGs Desa achievements, which also means the achievement of its truth, is best carried out with the census method. As data of all citizens, all families, and all areas of RTs is needed, the survey method is not appropriate, because it only takes sample data, examples, or just a few of the RTs, families, and village residents.

After determining the census method, then the data collection technique is more focused, specifically through questionnaires. Questions in the questionnaire can be in the form of closed or open questions. The enumerator must follow instructions and the order of the questions when asking the respondents for their responses. It is not allowed to guide respondents in their answers, and it is forbidden to explain at length the purpose of the question, because such things result in biased data acquisition.

Criticism of teleological epistemology could also be directed at SDGs Desa. Teleological criticism at the stages of modernisation development to the stages of society according to Marxism, is directed at the ethics of victims for the lower classes in order to achieve goals. There is a sacrifice to achieve the ultimate goal (*telos*) which should be carried out by the lower layers of society.

It seems inappropriate to aim such criticism at SDGs Desa as, since the beginning of the implementation of SDGs Desa, it has empowered the lowest classes, and improved the environment especially for groups that have been marginalised in the development process. Certainly, the process of reducing poverty to 0% fully benefits the lower classes. The process of achieving 100% secondary school participation also ensures children from poor families attend school. Disaster mitigation in all RTs protects all village residents, and disaster aid is specifically prioritised for the lower classes.

In everyday conversation today, it is impossible to deny the truth of the goals of the SDGs Desa. There are no village leaders who aim to increase or maintain the existence of poverty; all strive to realise a Village without Poverty (SDGs Desa goal1). There are no village leaders whose objective is to damage the terrestrial environment; on the contrary, the goal of building development is to realise a Village Caring for the Land Environment (SDGs Desa goal 15).

The history of the diversity of development paths to achieve similar development goals is shown in the difference in the flow of modernity throughout the first half of the twentieth century (Moore 1966). The path of economic democracy was taken by the United States and Britain; the path of communism was taken by the Soviet Union (which has now broken up); the path of fascism was taken by Germany, Japan, and Italy.

The diversity of development paths is conveyed to show that there has been a similar pattern evident in SDGs Desa, that is, the same goal is pursued with different paths. The SDGs Desa set the ultimate goal as the direction of village development. Each village can be creative, carrying out a variety of activities to achieve these goals.

3.3 Methodological Consequences of Interparty Cooperation

The structure of the SDGs Desa includes the citizenship dimension in goals 1, 2, 3, 4, 5, and 6. This means that the achievement of these goals is related to meeting the basic needs of villagers, namely freedom from poverty, freedom from hunger, health, education, and the involvement of village women.

The territorial dimension is spread across the SDGs Desa from goals 7–18. This means that the achievement of these goals is related to regional development, both in terms of settlements (SDGs Desa goal 11: Safe and Comfortable Village Residential Areas), neighbourhood areas (for example, SDGs Desa goal 9: Needs-based Village Infrastructure and Innovation), to the village area (for example, SDGs Desa goal 8: Equitable Village Economic Growth). Within the territorial dimension, there are aspects of the process and support for development in SDGs Desa goals 16 and 17, as well as the context of development at the village level in SDGs Desa goal 18.

From the beginning, the SDGs emphasised the aspect of stakeholder support for development. Specifically, this is reflected in goal 17 about cooperation, whether in the form of economic, social, and cultural cooperation networks between parties, or technical networks such as the internet. In addition, broadly, cooperation also includes achieving all the goals of the SDGs Desa. Various parties can work together to achieve SDGs Desa goal 7: Clean and Renewable Energy Village, for example, where the private sector provides solar power generators, which are trained by non-governmental organisations, and the village redeems them with village funds.

Networks also occur among the goals within the SDGs Desa. There is a connection between efforts to increase school access (SDGs Desa goal 4: Quality Village Education), with school participation for poor children (SDGs Desa goal 1: Village without Poverty), and school participation for girls (SDGs Desa goal 5: Involvement of Village Women).

The network of relationships between goals in the SDGs Desa has consequences for the methodology for data analysis. In addition to descriptive analysis to understand the specific conditions of the village, and the gap between the current conditions and the ultimate goal to be achieved in 2030, a network analysis is also needed that contains correlations across all SDGs Desa goals and their indicators.

There are two uses of correlation to measure the network of relationships between SDGs Desa. First, with specific data in a village, it can be shown that efforts to achieve certain SDGs Desa will be related to the results of achieving other SDGs Desa. Second, a SDGs Desa that has the most relationships and has the highest correlation value deserves to be prioritised for achievement. The first step to achieving the SDGs Desa can start with the results of data collection that show the goal that has the most relationships with other goals and has the highest relationship or correlation value.

The recapitulation of household data becomes village data, then becomes sub-district data that can also provide information on priority programmes and activities for the sub-district to build villages to achieve SDGs Desa. The accumulation of data and the results of sub-district analysis can be developed for the district/city level, then the province, and finally the national level.

3.4 Methodological Consequences of Village Level Application

The uniqueness of the SDGs Desa lies in goal 18: Dynamic Village Institutions and Adaptive Village Culture. Consequently, the epistemology of the SDGs Desa leads to efforts to achieve knowledge of the truth about a certain village. Explicitly, the methodology and measurement of SDGs Desa aim to ascertain the truth of a village under study. The boundary of study on one village becomes the delineation of the study context. Therefore, the measurement of SDGs Desa can be classified into village studies.

The context of the village has implications for the effort to collect microdata in order to be able to describe the condition of the village and explain various institutional dynamics and cultural adaptations within the village. At the citizen level, microdata takes the form of the conditions of each individual villager, covering aspects of citizenship within the SDGs Desa. At the family level, microdata also shows the condition of each family in the village, covering aspects of citizenship, as well as the influence of territorial conditions experienced by the family. At the RT level, microdata is aimed at digging up information about the territory closest to the family. Data at the village level is aimed at unearthing various aspects of village governance and the impact of development.

The methodology of the SDGs Desa is primarily aimed at measuring, explaining, understanding, and ultimately building a village. The participatory aspect is demonstrated by the preparation of questionnaire instruments in accordance with the needs of building a village, as shown by the results of previous qualitative and quantitative studies. The questionnaire opens up opportunities for updating, especially with added questions in accordance with new information needed by the village.

The village participates through funding with village funds or other legitimate budgets. A group of villagers and village associations are organised as the Village Data Collection Volunteer Team. They are the ones who survey their own village, survey the area of their RTs, survey families that have been their neighbours, as well as family members who are local villagers.

The questionnaire itself is a quantitative methodology instrument. The questionnaire is not used for surveys, but as an instrument for a census of village, RTs, family, and individuals. The census avoids the detached process of random respondent selection. What is implemented instead strengthens the validity of the data, by trained data collectors gathering information from villagers, families, RT environments, and the conditions of the village they are familiar with.

Another participatory aspect in the SDGs Desa census is the ownership of the village's personal data. The data is not owned by supra-village parties or parties above the village, but is owned by the village itself; its confidentiality is maintained, and it is used for the benefit of the village itself.

Pragmatic epistemology is demonstrated by the use of data in recommendations for village development. Instruments and data analysis are directed for use by the village in formulating annual development plans, six-year plans, and longer-term plans. The truth of the analysis is shown by its usefulness for building the village according to data and accelerating village development, so that the welfare of the villagers is soon realised.

References

Agusta, I. 2014. *Paradigm of social science methodology, crossroads of theory construction.* Bogor: IPB Press.

Creswell, J. W., and V. L. P. Clark. 2018. *Designing and implementing mixed methods research,* 2nd ed. Yogyakarta: Pustaka Pelajar.

Hunnex, M.D. 2004. *Philosophy map: Chronological and thematic approach.* Bandung: Teraju.

Moore, B., Jr. 1966. *Social origins of dictatorship and democracy: Lord and peasant in the making of modern world.* Boston: Beacon Press.

Chapter 4
Formulation Process Measurement

4.1 Academic Control

The control of SDGs Desa measurement from academics includes academics from Yogyakarta State University, UIN Sunan Kalijaga Yogyakarta, and Surabaya State University. Academic control takes the form of checking the framework of thought, concepts, operational definitions, and instruments. Academics also make recommendations for improvements to the drafts that have been prepared.

Academics also conduct an internal validity test of the instrument, by checking the suitability of SDGs Desa with the concept of SDGs at the global and national level. The results of the validity test by academics are declared valid, because they are based on the metadata of Presidential Regulation 59/2017 on the achievement of sustainable development goals (Bappenas 2020a, b, c, d), which is taken specifically on targets and indicators that are precisely built on the context of the village.

The SDGs at the global level are composed of 196 indicators (Fig. 4.1). When broken down at the national level in Presidential Regulation 59/2017, there are 241 indicators. Furthermore, its localisation in SDGs Desa includes 222 indicators. This consists of 210 global and national SDGs indicators that can be applied in villages, plus an additional 12 indicators for the 18th Goal of SDGs Desa, which are adjusted according to the results of village development research.

4.2 Field Data Collection Instrument Test

The localisation of indicators is compiled into a questionnaire, then field-tested. From 26–30 November 2020, a questionnaire test was conducted in Sumberagung Village, Perak Subdistrict, and Bawangan Village, Ploso Subdistrict, Jombang Regency, East Java.

A. H. Iskandar, *The Methods of Measuring the National Sustainable Development Goals in Indonesia*, Engaging Indonesia, https://doi.org/10.1007/978-981-96-0927-7_4

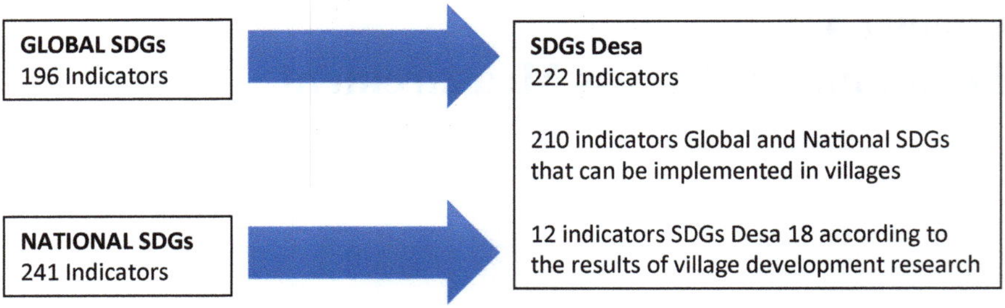

Fig. 4.1 Localisation of SDGs indicators from the global, national, to the village level

The number of questionnaires tested includes 2 village questionnaires, 18 RT questionnaires, 77 family questionnaires, and 216 individual questionnaires. There are only 2 village questionnaires corresponding to the number of test location villages, and it turns out that all questions can be answered by the village apparatus.

The reliability test of the questionnaire with Cronbach's alpha shows that the RT questionnaire reached a Cronbach's alpha value of 0.866. This means that 86.6 percent of the social reality level that the SDGs Desa expect can be drawn from the field. The family questionnaire reached a Cronbach's alpha value of 0.955. This means that 95.5 percent of the family level social reality expected by the SDGs Desa can be drawn from the field. Meanwhile, the individual questionnaire reached a Cronbach's alpha value of 0.731. This means that 73.1% of the individual level social reality expected by the SDGs Desa can be drawn from the field.

The validity test of the questionnaire was run with Pearson correlation, and directed at questions whose answers are interval and ratio values. The results show that for the RT questionnaire, all questions are valid. The family questionnaire shows that all questions in it are valid. As for the individual questionnaire, all questions are also valid.

Because the reliability level of the four questionnaires as data collection instruments has reached a number above 0.600 and can be answered by villagers, it is concluded that all SDGs Desa questionnaires are reliable. The validity of the questions in the four types of questionnaires is also valid.

Therefore, it is concluded that the SDGs Desa data collection instrument questionnaire is ready to be used for data collection at the individual or village resident level, families in the village, RT conditions, and village area conditions.

4.3 Case Study

The next step is to run a pilot study in December 2020. There are four different types of villages, and they are chosen to show that the SDGs Desa questionnaire can be used in various types of villages, with different ecologies, but a relatively similar number of families.

The type of village with a coastal ecology in the south of Java Island is represented by Kemojing Village, Binangun Subdistrict, Cilacap Regency, Central Java Province, inhabited by 744 families. For the type of village with a mountainous ecology in the central part of Java Island, Tempel Sari Village, Tretep Subdistrict, Temanggung Regency, Central Java Province, inhabited by 732 families, was chosen. The coastal ecology village in the north of Java Island selected Mlaten Village, Kalitidu Subdistrict, Bojonegoro Regency, East Java Province, which inhabited by 732 families. The mountainous ecological village type in the south of Java Island was chosen as Kretek Village, Taman Krocok Subdistrict, Bondowoso Regency, East Java Province. There are 746 families living in this village.

The case study was conducted in the form of a census to all residents, families, and RT in these villages. The results are in the form of village profiles, population profiles, citizenship issues, territorial issues, recommendations for problem-solving at the individual, family, regional levels. The data is also processed into the level of SDGs Desa achievement, and the rate of SDGs Desa target fulfilment.

References

Bappenas. 2020a. *Metadata indicator of economic development pillar*. Jakarta: Bappenas.
Bappenas. 2020b. *Metadata indicator of law and governance development pillar*. Jakarta: Bappenas.
Bappenas. 2020c. *Metadata indicator of environmental development pillar*. Jakarta: Bappenas.
Bappenas. 2020d. *Metadata indicator of social development pillar.* Jakarta: Bappenas.

Chapter 5
Conceptual Framework

The localisation of the SDGs concept at the global level and Sustainable Development Goals at the national level into SDGs Desa is done by localising metadata from the global and national levels to the village level. This localisation includes goals, targets, and indicators for measurement.

To facilitate readers' comprehension of the SDGs localization, Table 5.1 has been provided. In the table, it is shown that the localization of SDGs starts from the goal first. For example, the first global goal, is to end all forms of poverty everywhere. The formulation of this global goal is then localized at the village level to become Villages without Poverty.

The formulation of global targets has been translated into national targets, but localization of targets is still needed down to the village level. For example, goal 9 at the global level includes the target of developing quality, reliable, sustainable and resilient infrastructure, including regional and cross-border infrastructure, to support economic development and human well-being, with a focus on affordable and equitable access for all. This is translated at the national level to include the construction of toll roads. Because the construction of toll roads is outside the authority of the village, the localization of this target is adapted to the authority of the village. The target for SDGs Desa goal 9 is for village roads in good condition to reach 100%.

Measurement efforts are also localized to be used at the village level. For example, the electrification ratio to achieve goal 7 is measured by the number of families in a village using electricity divided by the total number of families in the village.

To emphasize the localization of SDGs into SDGs Desa, goal 18: Dynamic Village Institutions and Adaptive Village Culture, was also formulated. Goal 18 is stated in the targets related to the development of village institutions and village culture. The measurement of its achievements is also in accordance with the dynamics of local institutions and the results of local cultural adaptation to village development.

A. H. Iskandar, *The Methods of Measuring the National Sustainable Development Goals in Indonesia*, Engaging Indonesia, https://doi.org/10.1007/978-981-96-0927-7_5

Table 5.1 Localisation of Concept and Metadata of SDGs Measurement, Sustainable Development Goals, and SDGs Desa

No	Global goal	Global target	National target	Village goal	Village target year 2030	Measurement
1	End all forms of poverty everywhere	1. By 2030, reduce at least by half the proportion of men, women and children of all ages, living in poverty in all its dimensions, according to national definitions	1.1 Decrease in poverty rate	1. Village without Poverty	1.1 Poverty rate reaches 0%	Village poverty rate Percentage of population below the village poverty line Percentage of families owning family assets below 9 aspects
		2. Implement nationally appropriate social protection systems and measures for all, including the poorest, and by 2030 achieve substantial coverage for the poor and vulnerable	2.1 Increased percentage of population who are participants in health insurance through the National Social Security System in the Health Sector		2.1 Percentage of population who are participants in health insurance through The National Social Security System (SJSN) in the Health Sector reaching 100%	Percentage of SJSN participants in the village
			2.2 Increased percentage of poor and vulnerable disabled people who receive assistance in meeting basic needs		2.2 Percentage of poor disabled people who receive assistance in meeting basic needs increase year by year up to 100% Number of female-headed households (Perempuan Kepala Keluarga/ PEKKA) who receive assistance in meeting basic needs increases year by year up to 100%	Percentage of poor families with disabled members receiving government assistance
			2.3 Decrease in the number of extremely poor families receiving unconditional cash transfers		2.3 Number of poor families eligible to receive unconditional cash transfers increases year by year up to 100%	Change in the percentage of extremely poor families receiving unconditional cash transfers year by year

(continued)

Table 5.1 (continued)

No	Global goal	Global target	National target	Village goal	Village target year 2030	Measurement
			2.4 Increased Participation in the National Social Security System (Sistem Jaminan Sosial Nasional/SJSN) for Employment Sector		2.4 Percentage of Participation in the National Social Security System (Sistem Jaminan Sosial Nasional/SJSN) for Employment Sector reaches 100%	Percentage of participation in the National Social Security System (Sistem Jaminan Sosial Nasional/ SJSN) for Employment Sector 100%
		3. By 2030, ensure that all men and women, especially the poor and vulnerable, have equal rights to economic resources, as well as access to basic services, ownership and control over land and other forms of ownership, inheritance, natural resources, new technologies, and appropriate financial services, including microfinance	3.1 Increased coverage of childbirth in healthcare facilities for the lowest 40% income population		3.1 Coverage of childbirth in healthcare facilities for the lowest 40% income population reaches 100%	Coverage of childbirth in healthcare facilities for the lowest 40% of income population
			3.2 Increased coverage of complete basic immunisation for children aged 12–23 months for 40% of the lowest-income families		3.2 Coverage of complete basic immunisation for children aged 12–23 months for the bottom 40% of income-earning families reaches 100%	Coverage of complete basic immunisation for children aged 12–23 months for 40% of the lowest-income families
			3.3 Increased coverage of contraceptive use in all ways for women aged 15–49 years for 40% of the lowest income population		3.3 Coverage of contraceptive use in all ways for married productive-age women (aged 18–49 years) for the 40% of the lowest income population reaches 100%	Coverage of contraceptive use in all ways for women aged 15–49 years for 40% of the lowest-income population
			3.4 Increasing access to adequate drinking water for the lowest 40% income population		3.4 Access to adequate drinking water for the lowest 40% income population reaches 100%	Access to adequate drinking water for the lowest 40% income population
			3.5 Access to adequate sanitation for the for the lowest 40% income population reaches 100%		3.5 Access to adequate sanitation for the lowest 40% income population reaches 100%	Access to adequate sanitation for the lowest 40% income population

(continued)

Table 5.1 (continued)

No	Global goal	Global target	National target	Village goal	Village target year 2030	Measurement
			3.6 Increasing number of low-income households that can access decent housing		3.6 The number of low-income households that can access decent housing reaches 100%	The number of low-income households that can access decent housing
			3.7 Increasing the Gross Enrolment Rate for Primary School/Equivalent		3.7 The Gross Enrolment Rate for Primary School/Equivalent in the village reaches 100%	The Gross Enrolment Rate for High School/Vocational School/Equivalent in the village
			3.8 Increasing the Gross Enrolment Rate for Junior High School/Equivalent		3.8 The Gross Enrolment Rate for Junior High School/Equivalent in the village reaches 100%	The Gross Enrolment Rate for Junior High School/Equivalent in the village
			3.9 Increasing the Gross Enrolment Rate for High School/Vocational School/Equivalent		3.9 The Gross Enrolment Rate for High School/Vocational School/Equivalent in the village reaches 100%	The Gross Enrolment Rate for High School/Vocational School/Equivalent in the village
			3.10 Ownership of birth certificates for the bottom 40% income population		3.10 Ownership of birth certificates for the bottom 40% income population reaches 100%	Ownership of birth certificates for the bottom 40% income population
			3.11 Increasing access to lighting for the bottom 40% income population		3.11 Access to lighting for the bottom 40% income population reaches 100%	Access to lighting for the bottom 40% income population

(continued)

Table 5.1 (continued)

No	Global goal	Global target	National target	Village goal	Village target year 2030	Measurement
		4. By 2030, build the resilience of the poor and those in vulnerable situations, and reduce their vulnerability to climate-related extreme events and other economic, social, environmental, and disaster shocks	4.1 Increase in the number of disaster risk reduction reinforcement locations area		4.1 The number of disaster risk reduction reinforcement locations in hamlet (Rukun Tetangga/RT) reaches 100%	The number of disaster risk reduction reinforcement locations in hamlet (Rukun Tetangga/RT)
			4.2 Fulfilment of basic needs of social disaster victims		4.2 Fulfilment of basic needs for 100% of social disaster victims	Psychosocial support for social disaster victims
			4.3 Implementation of psychosocial support for social disaster victims		4.3 Psychosocial support for 100% of social disaster victims	Psychosocial support for social disaster victims
			4.4 Increase in the number of natural disaster/social disaster areas that receive special service education		4.4 100% of hamlets (Rukun Tetangga/RT) experiencing natural disasters/social disasters receive special services 100% of hamlets (Rukun Tetangga/RT) experiencing natural disasters/social disasters receive special services	100% of hamlets (Rukun Tetangga/RT) experiencing natural disasters/social disasters receive special services
			4.5 Decrease in disaster risk index at high-risk growth centres		4.5 Disaster reduction in hamlet (RT)	Disaster reduction in hamlet (RT)
			5.1 Improvement in the quality of priority development funding allocation in efforts to alleviate poverty, especially within the framework of micro policy through social and productive economic assistance		5.1 Increase in village financial allocation for poverty alleviation through social and productive economic assistance	Village financial allocation for poverty alleviation through social and productive economic assistance
			5.2 Improvement in the quality of priority development funding allocation related to the effectiveness of education budget utilisation, sustainability of funding and strengthening of social protection governance as well as increased health budget utilisation		5.2 Increase in village financial allocation for poverty alleviation through education assistance and social protection as well as health	Village financial allocation for poverty alleviation through assistance education and social protection as well as health

(continued)

Table 5.1 (continued)

No	Global goal	Global target	National target	Village goal	Village target year 2030	Measurement
2	II. Eliminating hunger, achieving food security and good nutrition, and promoting sustainable agriculture	1. By 2030, eliminate hunger and ensure access for all people, especially the poor and those in vulnerable conditions, including infants, to safe, nutritious, and sufficient food all year round	1.1 Decrease in the prevalence of malnutrition (underweight) in children under five	2. Village without Hunger	1.1 Prevalence of malnutrition (underweight) in children under five reaches 0%	Prevalence of malnutrition (underweight) in children under five
			1.2 Decrease in the proportion of the population with a minimum calorie intake below 1400 kcal/capita/day		1.2 Proportion of the population with a minimum calorie intake below 1400 kcal/capita/day reaches 0%	Proportion of the population with a minimum calorie intake below 1400 kcal/capita/day
		2. By 2030, eliminate all forms of malnutrition, including achieving internationally agreed targets for stunted and underweight children under 5 years of age by 2025, and meet the nutritional needs of adolescent girls, pregnant and lactating women, and the elderly	2.1 Decrease in the prevalence of stunting (short and very short) in children under two years of age		2.1 Prevalence of stunting (short and very short) in children under two years of age reaches 0%	Prevalence of stunting (short and very short) in children under two years of age
					2.2. Prevalence of adolescent girls experiencing anaemia reaches 0%	Percentage of adolescent girls experiencing anaemia
			2.2 Decrease in the prevalence of wasting (thinness) in children under five		2.2 Prevalence of wasting (thinness) in children under five reaches 0%	Prevalence of wasting (thinness) in children under five
			2.3 Decrease in the prevalence of anaemia in pregnant women		2.3 Prevalence of anaemia in pregnant women reaches 0%	Prevalence of anaemia in pregnant women
			2.4 Percentage of infants under 6 months receiving exclusive breastfeeding		2.4 Infants under 6 months receiving exclusive breastfeeding reaches 0%	Infants under 6 months receiving exclusive breastfeeding
			2.5 Improvement in the quality of food consumption indicated by the score of the Expected Food Pattern (Pola Pangan Harapan/PPH) reaching 92.5 (2014: 81.8), and the level of fish consumption becoming 54.5 kg/capita/year		2.5 Quality of food consumption indicated by the score of the Expected Food Pattern (Pola Pangan Harapan/PPH) reaches 100	Quality of food consumption indicated by the score of the Expected Food Pattern (Pola Pangan Harapan/PPH)

(continued)

Table 5.1 (continued)

No	Global goal	Global target	National target	Village goal	Village target year 2030	Measurement
		3. By 2030, ensure sustainable food production systems and implement resilient agricultural practices that increase production and productivity, help maintain ecosystems, strengthen the adaptive capacity to climate change, extreme weather, drought, floods, and other disasters, and progressively improve the quality of soil and land	3.1 Increased added value per agricultural worker		3.1 Increased income of farmers and farm labourers	Total income of farmers and farm labourers
			3.2 Protection of paddy fields designated as Sustainable Food Agriculture Land (Lahan Pertanian Pangan Berkelanjutan/ LP2B) becomes 100%		3.2 There are sustainable food agriculture areas	There are sustainable food agriculture areas
3	III. Ensuring healthy lives and promoting well-being for all at all ages	1. By 2030, reduce the maternal mortality ratio to less than 70 per 100,000 live births	1.1 Decrease in maternal mortality rate per 100,000 live births	3. Healthy and Prosperous Village	1.1 Maternal mortality rate per 100,000 live births reaches 0	Maternal mortality rate per 100,000 live births
			1.2 Increased percentage of births in health care facilities		1.2 Percentage of births at health facilities reaches 100%	Percentage of births at health facilities
			1.3 Increased percentage of births by skilled health personnel		1.3 Percentage of births by skilled health personnel reaches 100%	Percentage of births by skilled health personnel
		2. By 2030, end preventable newborn and child deaths, with all countries striving to reduce Neonatal Mortality Rate at least to 12 per 1,000 live births and Child Mortality Rate to 25 per 1,000	2.1 Decrease in infant mortality rate per 1000 live births in 2019		2.1 Infant mortality rate per 1000 live births reaches 0	Infant mortality rate per 1000 live births
			2.2 Increased percentage of districts/ cities achieving 80% complete basic immunisation in infants		2.2 Percentage of hamlets (Rukun Tetangga/RT) with complete basic immunisation in infants reaches 100%	Percentage of hamlets (Rukun Tetangga/RT) with complete basic immunisation in infants
		3. By 2030, end the epidemics of AIDS, tuberculosis, malaria, and neglected tropical diseases, and combating hepatitis, waterborne diseases, and other infectious diseases	3.1 Decrease in HIV prevalence in the adult population		3.1 HIV prevalence in the adult population reaches 0%	HIV prevalence in the adult population
			3.2 Decrease in Tuberculosis (TB) prevalence per 100,000 population		3.2 Tuberculosis (TB) prevalence per 100,000 population reaches 0%	Tuberculosis (TB) prevalence per 100,000 population

(continued)

Table 5.1 (continued)

No	Global goal	Global target	National target	Village goal	Village target year 2030	Measurement
			3.3 Increase in the number of districts/ cities with malaria elimination		3.3 The number of households with malaria elimination reaches 100%	Number of households with malaria elimination
			3.4 Increase in the number of provinces with leprosy elimination		3.4 The number of households with leprosy elimination reaches 100%	Number of households with leprosy elimination
			3.5 Increase in the number of districts/ cities with filariasis elimination		3.5 The number of hamlets (Rukun Tetangga/RTs) with filariasis (elephantiasis) elimination reaches 100%	Number of hamlets (Rukun Tetangga/ RTs) with filariasis (elephantiasis) elimination
		4. By 2030, reduce by one third premature mortality from non-communicable diseases through prevention and treatment, and promote mental health and well-being	4.1 Decrease in the percentage of smoking in the population aged ≤ 18 years		4.1 The percentage of smoking in the population aged ≤ 18 years reaches 0%	Percentage of smoking in the population aged ≤ 18 years
			4.2 Decrease in the prevalence of high blood pressure		4.2 The prevalence of high blood pressure reaches 0%	Prevalence of high blood pressure
			4.3 No increase in the prevalence of obesity in the population aged 18 years and over		4.3 The prevalence of obesity in the population aged 18 years and over reaches 0%	Prevalence of obesity in the population aged 18 years and over
			4.4 Increase in the number of Districts/ Cities that have health centres that conduct mental health efforts		4.4 Increase in the number of RTs that have integrated health posts that conduct mental health efforts	Number of RTs that have integrated health posts that conduct mental health efforts

(continued)

Table 5.1 (continued)

No	Global goal	Global target	National target	Village goal	Village target year 2030	Measurement
		5. Strengthen the prevention and treatment of substance abuse, including drug abuse and harmful alcohol use	5.1 Increase in the number of victims of narcotics, psychotropics and other addictive substances (narkotika, psikotropika dan zat adiktif lainnya/ NAPZA) abuse receiving social rehabilitation in homes according to service standards		5.1 The number of victims of narcotics, psychotropics and other addictive substances (narkotika, psikotropika dan zat adiktif lainnya/ NAPZA) abuse receiving social rehabilitation in homes according to service standards reaches 100%	The number of victims of narcotics, psychotropics and other addictive substances (narkotika, psikotropika dan zat adiktif lainnya/ NAPZA) abuse receiving social rehabilitation in homes according to service standards
			5.2 Increase in the number of Social Rehabilitation Institutions for victims of narcotics, psychotropics and other addictive substances (narkotika, psikotropika dan zat adiktif lainnya/ NAPZA) abuse that have been developed/assisted			
			5.3 Control of the prevalence rate of drug abuse		5.3 Prevalence of drug abuse reaches 0%	Prevalence of drug abuse
		6. By 2030, reduce by half the global number of deaths and injuries from traffic accidents	6.1 Decrease in the number of deaths and serious injuries due to traffic accidents		6.1 The number of deaths and serious injuries due to traffic accidents reaches 0%	The number of deaths and serious injuries due to traffic accidents
		7. By 2030, ensure universal access to sexual and reproductive health services, including family planning, information and education, and the integration of reproductive health into national strategies and programmes	7.1 Decrease in the Total Fertility Rate (TFR)		7.1 Total Fertility Rate (TFR) below 1.5	Total Fertility Rate (TFR)
			7.2 Increase in the prevalence of short-term contraceptive use		7.2 Prevalence of short-term contraceptive use in married people of productive age (18–49 years old) reaches 100%	Prevalence of short-term contraceptive use in married people of productive age (18–49 years old)

(continued)

Table 5.1 (continued)

No	Global goal	Global target	National target	Village goal	Village target year 2030	Measurement
			7.3 Increase in the use of long-term contraceptive methods (modern method)		7.3 The use of long-term contraceptive methods (modern method) in married people of productive age (18–49 years old) reaches 100%	The use of long-term contraceptive methods (modern method) in married people of productive age (18–49 years old)
			7.4 Decrease in the birth rate in teenagers aged 15–19 years (age specific fertility rate/ASFR)		7.4 The birth rate in teenagers aged 15–19 years (age specific fertility rate/ASFR) reaches 0%	The birth rate in teenagers aged 15–19 years (age specific fertility rate/ASFR)
		8. Achieve universal health coverage, including financial risk protection, access to good basic health services, and access to safe, effective, quality, and affordable basic medicines and vaccines for everyone	8.1 Decrease in unmet need for health services		8.1 Unmet need for health services reaches 0%	Unmet need for health services
			8.2 Increase in National Health Insurance (JKN) coverage		8.2 National Health Insurance (Jaminan Kesehatan Nasional/JKN) coverage reaches 100%	National Health Insurance (Jaminan Kesehatan Nasional/JKN) coverage
4	IV. Ensuring inclusive and equitable quality education and promoting lifelong learning opportunities for all	1. By 2030, ensure that all girls and boys complete free, equitable and quality primary and secondary education leading to relevant and effective learning outcomes	1.1 Increase in the percentage of primary school/equivalent (SD/MI) accredited at least B	4. Quality Village Education	1.1 Percentage of access to primary school/equivalent (SD/MI) accredited at least B reaches 100%	Percentage of access to primary school/equivalent (SD/MI) accredited at least B
			1.2 Increase in the percentage of junior high school/equivalent (SMP/MTs) accredited at least B		1.2 Percentage of access to junior high school/equivalent (SMP/MTs) accredited at least B reaches 100%	Percentage of access to junior high school/equivalent (SMP/MTs) accredited at least B
			1.3 Increase in the percentage of senior high school/equivalent (SMA/MA) accredited at least B		1.3 Percentage of access to senior high school/equivalent (SMA/MA) accredited at least B reaches 100%	Percentage of access to senior high school/equivalent (SMA/MA) accredited at least B

(continued)

Table 5.1 (continued)

No	Global goal	Global target	National target	Village goal	Village target year 2030	Measurement
					1.4 Increase in Gross Enrolment Rate (GER) for primary school/ equivalent (SD/MI)	Percentage of access to primary school/ equivalent/Islamic boarding schools (pesantren)
			1.4 Increase in Gross Enrolment Rate (GER) for primary school/equivalent (SD/MI)		1.4 Gross Enrolment Rate (GER) for primary school/equivalent (SD/ MI) reaches 100%	Gross Enrolment Rate (GER) for primary school/ equivalent (SD/MI)
			1.5 Increase in GER for junior high school/equivalent (SMP/MTs)		1.5 GER for junior high school/equivalent (SMP/MTs) reaches 100%	GER for junior high school/equivalent (SMP/MTs)
			1.6 Increase in GER for senior high school/senior vocational high school/ senior Islamic high school/equivalent (SMA/SMK/MA)		1.6 GER for senior high school/senior vocational high school/ senior Islamic high school/equivalent (SMA/SMK/MA) reaches 100%	GER for senior high school/senior vocational high school/senior Islamic high school/ equivalent (SMA/ SMK/MA)
			1.7 The average length of schooling for people aged over 15 years old increased in 2019 to 8.8 years (2015: 8.25 years)		1.7 The average length of schooling for people aged over 15 years old reaches 12 years	Average length of schooling for people aged over 15 years old
		2. By 2030, ensure that all girls and boys have access to quality early childhood development, care, and pre-primary education so that they are ready for primary education	2.1 Increase in the Gross Enrolment Rate (GER) of children attending Early Childhood Education (Pendidikan Anak Usia Dini/PAUD)		2.1 GER of children attending Early Childhood Education (Pendidikan Anak Usia Dini/PAUD) and Kindergarten (Taman Kanak-kanak) reaches 100%	GER of children attending Early Childhood Education (Pendidikan Anak Usia Dini/PAUD) and Kindergarten (Taman Kanak-kanak)
		3. By 2030, ensure equal access for all women and men to affordable and quality technical, vocational and higher education, including university	3.1 Increase in the GER of senior high school/senior vocational high school/ equivalent		3.1 GER of senior high school/senior vocational high school/ equivalent reaches 100%	GER of senior high school/senior vocational high school/equivalent
			3.2 Increase in the GER of Higher Education (HE)			

(continued)

Table 5.1 (continued)

No	Global goal	Global target	National target	Village goal	Village target year 2030	Measurement
		4. By 2030, eliminate gender disparities in education and ensure equal access to all levels of education and vocational training for vulnerable populations including persons with disabilities, indigenous peoples, and children in vulnerable conditions	4.1 Ratio of the Net Enrolment Rate (NER) of females/males in primary school/equivalent/Package A that is gender equal		4.1 Ratio of the NER of females/males in primary school/equivalent/Package A that is gender equal reaches 100%	Ratio of the NER of females/males in primary school/equivalent/Package A that is gender equal
			4.2 Ratio of NER of females/males in junior high school/equivalent/Package B that is gender equal		4.2 Ratio of NER of females/males in junior high school/equivalent/Package B that is gender equal reaches 100%	Ratio of NER of females/males in junior high school/equivalent/Package B that is gender equal
			4.3 Ratio of GER of females/males in senior high school/senior vocational high school/equivalent that is gender equal		4.3 Ratio of GER of females/males in senior high school/senior vocational high school/equivalent that is gender equal reaches 100%	Ratio of GER of females/males in senior high school/senior vocational high school/equivalent that is gender equal
			4.4 Ratio of GER of females/males in HE and PTA that is gender equal			
		5. By 2030, ensure that all adolescents and a certain proportion of adult groups, both men and women, have literacy and numeracy skills	5.1 Increase in the average literacy rate of the population aged over 15 years		5.1The literacy rate in Latin and non-Latin for the population aged over 15 years reaches 100%	The literacy rate in Latin and non-Latin for the population aged over 15 years
			5.2 Increase in the percentage of the literacy rate of the adult population aged 15–59 years		5.2 The literacy rate in Latin and non-Latin for the adult population aged 15–59 years reaches 100%	The literacy rate in Latin and non-Latin for the adult population aged 15–59 years
					5.3 Availability of Community Reading Parks (Taman Bacaan Masyarakat/TBM) or libraries	Availability of Community Reading Parks (Taman Bacaan Masyarakat/TBM) or libraries

(continued)

Table 5.1 (continued)

No	Global goal	Global target	National target	Village goal	Village target year 2030	Measurement
5	V. Achieve gender equality and empower women	1. End all forms of discrimination against women everywhere	1.1 Increase in the number of gender-responsive policies supporting women's empowerment	5. Involvement of Village Women	1.1 Gender-responsive policies supporting women's empowerment at least 30%	Gender-responsive policies supporting women's empowerment
		2. Eliminate all forms of violence against women in public and private spaces, including human trafficking and sexual exploitation, as well as various other types of exploitation	2.1 Decrease in the prevalence of cases of violence against women		2.1 Prevalence of cases of violence against women reaches 0%	Prevalence of cases of violence against women
			2.2 Increase in the percentage of cases of violence against women who receive comprehensive services		2.2 Cases of violence against women who receive comprehensive services reach 100%	Cases of violence against women who receive comprehensive services
		3. Eliminate all harmful practices, such as child marriage, early and forced marriage, and female circumcision	3.1 Increase in the median age of first marriage for women (maturation of the age of first marriage)		3.1 Median age of first marriage for women (maturation of the age of first marriage) above 18 years	Median age of first marriage for women (maturation of the age of first marriage)
			3.2 Decrease in ASFR 15–19 years		3.2 ASFR 15–19 years below 1.5	ASFR 15–19 years
			3.3 Increase in GER of senior high school/senior vocational school/equivalent		3.3 GER of senior high school/senior vocational school/equivalent reaches 100%	GER of senior high school/senior vocational school/equivalent
		4. Ensure full and effective participation, and equal opportunities for women to lead at all levels of decision-making in political, economic, and societal life	4.1 Increase in women's representation in House of Representatives		4.1 Percentage of women in the Village Consultative Body (Badan Permusyawaratan Desa/BPD) at least 30%	Percentage of women in the Village Consultative Body (BPD)
			4.2 Increase in women's representation as decision-makers in executive institutions (Echelon I and II))		4.2 Percentage of women as village officials at least 30%	Percentage of women as village officials

(continued)

Table 5.1 (continued)

No	Global goal	Global target	National target	Village goal	Village target year 2030	Measurement
					4.3 Percentage of women attending village deliberation forum at least 30%	Percentage of women attending village deliberation forum
					4.4 Percentage of women participating in village development activities at least 30%	Percentage of women participating in village development activities
		5. Ensure universal access to sexual and reproductive health, and reproductive rights as agreed in accordance with the Programme of Action of the International Conference on Population and Development and the Beijing Platform and the review outcome documents of these conferences.	5.1 Decrease in unmet need for family planning		5.1 Unmet need for family planning reaches 0%	Unmet need for family planning
			5.2 Increase in knowledge and understanding of Couples of Childbearing Age (Pasangan Usian Subur/PUS) about at least 4 types of modern contraception methods		5.2 Couples of Childbearing Age (Pasangan Usia Subur/ PUS) understand at least 4 types of modern contraception methods	Couples of Childbearing Age (Pasangan Usia Subur/PUS) who understand at least 4 types of modern contraception methods
			5.3 Availability of regulations that ensure women to get services, information, and education related to family planning and reproductive health		5.3 There are regulations that ensure women to get services, information, and education related to family planning and reproductive health	Regulations that ensure women to get services, information, and education related to family planning and reproductive health
6	VI. Ensuring the availability and sustainable management of clean water and sanitation for all	1. By 2030, achieve universal and equitable access to safe and affordable drinking water for all	1.1 Increased access to adequate drinking water services	6. Village with Clean Water and Sanitation	1.1 Access to adequate drinking water services reaches 100% of families	Access to adequate drinking water services for families
			1.2 Increased capacity of raw water infrastructure to serve households, urban and industrial		1.2 Families and industries served by raw water reach 100%	Families and industries served by raw water

(continued)

Table 5.1 (continued)

No	Global goal	Global target	National target	Village goal	Village target year 2030	Measurement
		2. By 2030, achieve access to adequate and equitable sanitation and hygiene for all, and end open defecation, paying special attention to the needs of women and vulnerable groups	2.1 Increased access to adequate sanitation		2.1 Family access to adequate sanitation reaches 100%	Family access to adequate sanitation
			2.2 Increased number of villages/ neighbourhoods implementing Community-Led Total Sanitation (Sanitasi Total Berbasis Masyarakat/ STBM)		2.2 Families implementing Community-Led Total Sanitation (Sanitasi Total Berbasis Masyarakat/STBM) reach 100%	Families implementing Community-Led Total Sanitation (Sanitasi Total Berbasis Masyarakat/ STBM)
		3. By 2030, improve water quality by reducing pollution, eliminating dumping, and minimising the release of hazardous materials and chemicals, halve the proportion of untreated wastewater, and significantly increase recycling and safe reuse globally	3.1 Establishment of wastewater infrastructure with a centralised system at the city, area, communal scale		3.1 Families and industries using wastewater facilities reach 100%	Families and industries using wastewater facilities
			3.2 Improvement of the quality of local wastewater management through the improvement of urban faecal sludge management and the construction of Faecal Sludge Treatment Installation (Instalasi Pengolahan Lumpur Tinja/ IPLT)		3.2 Families and industries using faecal sludge management facilities reach 100%	Families and industries using faecal sludge management facilities
			3.3 Water quality management, both in rivers, reservoirs, lakes, river mouths, beaches including the improvement of hydrological and water quality monitoring systems with the indicator of improving water quality		3.3 The quality of the water source in hamlet (Rukun Tetangga/RT) is colourless, odourless, tasteless	The quality of the water source in hamlet (Rukun Tetangga/RT) is colourless, odourless, tasteless
			3.4 Improvement of river water quality as a raw water source towards the average quality standard of class II river water			
		4. By 2030, significantly increase water use efficiency across all sectors, and ensure the use and supply of fresh water is sustainable to address water scarcity and significantly reduce the number of people suffering from water scarcity	4.1 Control and law enforcement for excessive groundwater use accompanied by acceleration of provision and management of raw water in economic areas, and the implementation of competitive industrial water tariff policies		4.1 There is a regulation on groundwater use	Regulation on groundwater use

(continued)

Table 5.1 (continued)

No	Global goal	Global target	National target	Village goal	Village target year 2030	Measurement
			4.2 Provide incentives for water conservation in agriculture/plantations and industry including the application of the reduce principle, developing reuse and recycle, and the development of the concept of safe use of wastewater in agriculture		4.2 Reduction of irrigation water costs in the industry for those implementing safe use of wastewater in agriculture	Reduction of irrigation water costs in the industry for those implementing safe use of wastewater in agriculture
		5. By 2030, implement integrated water resource management at all levels, including through appropriate cross-border cooperation	5.1 Internalisation of 108 Integrated River Basin Management Plans that have been prepared into the Regional Spatial Plan		5.1 Availability of environmental preservation regulations around river flows	Availability of environmental preservation regulations around river flows
			5.2 Updating and revitalising hydrological and climatological stations in 8 River Regions and Establishment of a water resource information network in 8 River Regions		5.2 Availability of information from the nearest hydrological and climatological stations	Information from the nearest hydrological and climatological stations
			5.3 Establishment of a water resource information network in 8 River Regions		5.3 Availability of water resource information	Water resource information
			5.4 Increase in the number of watersheds (Daerah Aliran Sungai/DAS) that have increased their number of springs and 19 watershed (Daerah Aliran Sungai/DAS) that have a Memorandum of Understanding (MoU) across countries in 2019		5.4 The number of springs remains	The number of springs remains
			5.5 Health recovery of watersheds through the development of Community Forest Plantations (hutan Tanaman Rakyat/HTR), Community Forests (Hutan Kemasyarakatan/HKm), Village Forests (Hutan Desa/HD), Customary Forests and People's Forests (Hutan Rakyat/HR) and the enhancement of Non-Timber Forest Products (Hasil Hutan Bukan Kayu/HHBK)		5.5 There are tree planting activities around the river flow	Tree planting activities around the river flow

(continued)

Table 5.1 (continued)

No	Global goal	Global target	National target	Village goal	Village target year 2030	Measurement
			5.6 Increasing community participation in the management of river and lake catchment areas		5.6 There are communal activities for the preservation of river and lake flows	Communal activities for the preservation of river and lake flows
			5.7 Continuing the institutional arrangement of water resources, among others by: Synergising the regulation of authority and responsibility at all levels of government along with all stakeholders and implementing it consistently; Improving communication, cooperation, and coordination between institutions and between coordination bodies for water resource management that have been formed; and Improving the institutional capacity of water resource management		5.7 There are regulations for water resource management Implementation of training/socialisation of water resource management	Regulations for water resource management Implementation of training/socialisation of water resource management
		6. By 2020, protect and restore ecosystems related to water resources, including mountains, forests, wetlands, rivers, groundwater, and lakes.	6.1 Improved water quality in the lake		6.1 Lake water quality is colourless, odourless, tasteless	Lake water quality is colourless, odourless, tasteless
			6.2 Increased lakes with less than 1% siltation		6.2 Implementation of annual lake dredging activities	Implementation of annual lake dredging activities
			6.3 Increased lakes with reduced erosion levels		6.3 No annual lake erosion events	No annual lake erosion events
			6.4 Reducing the extent of critical land through rehabilitation within the Forest Management Unit (Kesatuan Pengelolaan Hutan/KPH)		6.4 No barren/bald land	No barren/bald land
			6.5 Protection of springs and Recovery of river health in 5 Priority Watersheds		6.5 The number of springs remains the same	The number of springs remains the same

(continued)

Table 5.1 (continued)

No	Global goal	Global target	National target	Village goal	Village target year 2030	Measurement
7	VII. Ensure affordable, reliable, sustainable, and modern energy access for all	1. By 2030, ensure universal access to affordable, reliable, and modern energy services	1.1 Increase in electrification ratio	7. Clean and Renewable Energy Village	1.1 The number of families using electricity reaches 100%	Number of families using electricity
			1.2 Increase in per capita electricity consumption to 1,200 KWh		1.2 Per capita electricity consumption rate over 1,200 KWh	Per capita electricity consumption rate
			1.3 Achievement of 1.1 million household gas connections		1.3 Families using gas or wood waste for cooking reach 100%	Families using gas or wood waste for cooking
		2. By 2030, substantially increase the proportion of renewable energy in the global energy mix	2.1 Renewable energy mix reaches 10–16% in 2019		2.1 Use of renewable energy mix reaches 60% of families	Use of renewable energy mix in families
		3. By 2030, double the global rate of improvement in energy efficiency	3.1 Primary energy intensity (decrease of 1% per year)		3.1 Reduction of families using oil for transportation and cooking reaches 50%	Reduction of families using oil for transportation and cooking
8	VIII. Promote inclusive and sustainable economic growth, productive and comprehensive job opportunities and decent work for all	1. Maintain per capita economic growth in accordance with national conditions and, in particular, at least 7% gross domestic product growth per year in less developed countries	1.1 Increase in Gross Domestic Product (GDP) per capita per year to more than Rp 50 million in 2019	8. Equitable Village Economic Growth	1.1 Average Village GDP per capita above Rp30 million	Village GDP per capita
		2. Achieve higher levels of economic productivity through diversification, upgrading and technological innovation, including through focus on sectors that provide high value-added and labour-intensive	2.1 Real GDP growth per working person increases		2.1. Average per capita Village GDP above Rp30 million per worker in the village	Village GDP per capita/worker in the village

(continued)

Table 5.1 (continued)

No	Global goal	Global target	National target	Village goal	Village target year 2030	Measurement
		3. Encourage development policies that support productive activities, the creation of decent work, entrepreneurship, creativity and innovation, and encourage the formalisation and growth of micro, small, and medium enterprises, including through access to financial services	3.1 Percentage of formal workforce reaches 51%		3.1 Percentage of workers in the formal sector reaches 51%	Percentage of workers in the formal sector
			3.2 Access to formal Financial Services for Micro, Small and Medium Enterprises (MSMEs) 25%		3.2 At least 51% of Micro, Small and Medium Enterprises (MSMEs) in the village have access to formal capital	Micro, Small and Medium Enterprises (MSMEs) in the village that have access to formal capital
		4. By 2030, achieve full and productive employment and decent work for all women and men, including for young people and persons with disabilities, and equal pay for work of equal value	4.1 Creation of job opportunities		4.1 Number of people employed reaches 100% of the workforce Unemployment rate in the village reaches 0%	Number of people employed in the village workforce Unemployment rate in the village
					4.2 Labour absorption in Cash-for-Work (Padat Karya Tunai Desa/PKTD: manpower-intensive projects financed in cash by village funds) reaches more than 50% of the unemployed in the village	Labour absorption in labour-intensive work in the village
		5. By 2020, substantially reduce the proportion of young people not in employment, education or training	5.1 Increase the skills of vulnerable workers so they can enter the labour market		5.1 New workforce members who have not worked and receive skills training reaches 100%	New workforce members who have not worked and receive skills training
		6. Take immediate and action to eradicate forced labour, end slavery and human trafficking, secure the prohibition and elimination of the worst forms of child labour, including recruitment and use of child soldiers, and by 2025 end child labour in all its forms	6.1 Strengthening of the child protection system including efforts to protect children from violence, exploitation, neglect, and other forms of maltreatment		6.1 Acts of violence, exploitation, neglect, and other forms of maltreatment of children reach 0%	Acts of violence, exploitation, neglect, and other forms of maltreatment of children

(continued)

Table 5.1 (continued)

No	Global goal	Global target	National target	Village goal	Village target year 2030	Measurement
		7. Protecting the rights of workers and promoting a safe and secure working environment for all workers, including migrant workers, especially female migrant workers, and those working in hazardous jobs	7.1 Creation of a safe and healthy working environment		7.1 Workplaces with health and safety facilities reach 100%	Workplaces with health and safety facilities
		8. By 2030, formulate and implement policies to promote sustainable tourism that creates jobs and promotes local culture and products	8.1 Increased contribution of tourism to 8% of GDP		8.1 The contribution of tourism activities reaches at least 8% of the Village GDP	Contribution of tourism activities to the Village GDP
			8.2 Increased number of foreign tourists		8.2 The number of tourists increases each year	Number of tourists each year
		9. Strengthen the capacity of domestic financial institutions to encourage and expand access to banking, insurance and financial services for all	9.1 Increased expansion of access to capital and financial services through strengthening financial services		9.1 There is access to capital and financial services	Access to capital and financial services
9	IX. Building resilient infrastructure, promoting inclusive and sustainable industries, and encouraging innovation	1. Developing quality, reliable, sustainable and resilient infrastructure, including regional and cross border infrastructure, to support economic development and human well-being, with a focus on affordable and equitable access for all.	1.1 Construction of toll roads	9. Needs-based Village Infrastructure and Innovation	1.1 Percentage of roads in good condition reaches 100%	Percentage of roads in good condition
			1.2 Increased number of ferry docks		1.2 Percentage of docks/boat moorings in good condition reaches 100%	Percentage of docks/ boat moorings in good condition
			1.3 Increase in the length of railway lines			
			1.4 Increase in the number of airports			
			1.5 Construction of strategic ports to support the sea toll			

(continued)

Table 5.1 (continued)

No	Global goal	Global target	National target	Village goal	Village target year 2030	Measurement
		2. Promoting inclusive and sustainable industrialisation, and by 2030, significantly increasing the proportion of industry in employment and gross domestic product, in line with national conditions, and doubling its proportion in less developed countries	2.1 Increasing the growth rate of the manufacturing industry GDP so it is higher than the GDP growth		2.1 The growth rate of the household, small and medium industries is faster than the Village GDP	The growth rate of the household, small and medium industries compared to the Village GDP
			2.2 Increasing the contribution of the manufacturing industry		2.2 The contribution of processing industry activities reaches at least 8% of the Village GDP	The contribution of processing industry activities to the Village GDP
		3. By 2030, enhance infrastructure and retrofit industries to be sustainable, with increased resource efficiency and better adoption of clean and environmentally friendly industrial technologies and processes, implemented by all countries according to their abilities	3.1 Reduction of CO_2 emissions		3.1 The percentage of industries polluting the air reaches 0%	The percentage of industries polluting the air
10	X. Reducing intra and inter-country disparities	1. By 2030, progressively achieve and sustain income growth of the population that is below 40% of the population at a higher level than the national average	1.1 Gini coefficient decreases	10. Village Without Gaps	1.1 Gini coefficient below 0.200	Village Gini coefficient
			1.2 Poverty rate decreases		1.2 Poverty rate 0%	Village poverty rate
			1.3 Increase in the number of uplifted disadvantaged areas			
			1.4 Reduction of Disadvantaged Villages			
			1.5 Increase in Self-Sufficient Villages		1.5 Self-sufficient village status	Village Development Status
			1.6 Decrease in the percentage of poor population in disadvantaged areas			

(continued)

Table 5.1 (continued)

No	Global goal	Global target	National target	Village goal	Village target year 2030	Measurement
		2. Ensuring equal opportunities and reducing outcome disparities, including by abolishing discriminatory laws, policies and practices, and promoting appropriate legislation, policies and actions related to such legislation and policies	2.1 Increase in Civil Liberty Index		2.1 Civil Liberty Index reaches 100	Civil Liberty Index
		3. Adopting policies, especially fiscal, wage and social protection policies, and progressively achieving greater equality	3.1 Increase in the participation of the National Social Security System in the employment sector for formal and informal workers		3.1 The number of workers participating in the National Social Security System for employment sector reaches 100%	Number of workers participating in the National Social Security System for employment sector
		4. Facilitating orderly, safe, regular and responsible human migration and mobility, including through the implementation of well-planned and managed migration policies	4.1 Increase in cooperation to protect the rights and safety of migrant workers		4.1 Violence against migrant workers reaches 0%	Violence against migrant workers
			4.2 Improvement in the governance of migrant worker placement		4.2 Availability of advocacy rules for prospective and migrant workers	Advocacy rules for prospective and migrant workers
11	XI. Making cities and human settlements inclusive, safe, resilient, and sustainable	1. By 2030, ensure access for all to adequate, safe, affordable housing, including slum area management, as well as access to basic urban services	1.1 Availability of access for households to adequate and affordable housing	11. Safe and Comfortable Village Residential Areas	1.1 Slum house percentage reaches 0%	Percentage of slum houses
			1.2 Realisation of meeting urban service standards in a city that is safe, comfortable and habitable in the aspect of settlement in the metropolitan city		1.2 Implementation of security in the hamlets (Rukun Tetangga/RT) reaches 100%	Security in the hamlets (Rukun Tetangga/RT)
			1.3 Realisation of meeting urban service standards in a city that is safe, comfortable and habitable in the aspect of settlement in Medium and New Cities			

(continued)

Table 5.1 (continued)

No	Global goal	Global target	National target	Village goal	Village target year 2030	Measurement
		2. By 2030, provide access to safe, affordable, accessible, and sustainable transport systems for all, improving road safety, notably by expanding public transport, with special attention to the needs of those in vulnerable situations: women, children, persons with disabilities, and the elderly	2.1 Increase in the share of public transportation users in urban areas		2.1 Number of families using Public transport reaches more than 50%	Families using public transportation
					2.2 Number of children, women, people with disabilities and elderly people who user public transportation reaches more than 50%	Percentage of children, women, people with disabilities, elderly users of public transportation
			2.2 Development of the railway transport system			
		3. By 2030, strengthen inclusive and sustainable urbanisation, as well as participatory capacity, planning for handling sustainable and integrated settlements in all countries	3.1 Optimisation of medium cities outside Java that are directed as controllers (buffers) of urbanisation flow and as main growth centres that drive the linkage of cities and villages		3.1 Population moving to the city reaches less than 15%	Percentage of village population moving to the city
			3.2 The establishment of 5 new Metropolitan areas outside Java as National Activity Centres			

(continued)

Table 5.1 (continued)

No	Global goal	Global target	National target	Village goal	Village target year 2030	Measurement
			3.3 Increased role of the private sector, community organisations and professional organisations actively, in the Sustainable City Planning and Development Dialogue Forum		There is involvement from private sector roles, community organizations, and professional organizations in the village activity of gathering together to chat, discuss, and socialise known as Cangkrukan, Jagongan, or Kongkow	There is involvement from private sector roles, community organizations, and professional organizations in the village activity of gathering together to chat, discuss, and socialise known as Cangkrukan, Jagongan, or Kongkow
			3.4 Availability of infrastructure financing institutions			
		4. Strengthen efforts to protect and preserve world cultural heritage and natural heritage	4.1 The realisation of liveable cities and urban areas through the development of heritage cities based on social cultural character in metropolitan urban areas, large, medium, and small cities		4.1 100% of culture is preserved	Percentage of culture preserved
		5. By 2030, significantly reduce the number of deaths and the number of people affected and substantially reduce economic losses relative to global GDP caused by disasters, including water-related disasters, with a focus on protecting the poor and people in vulnerable situations	5.1 Decrease in Disaster Risk Index (Indeks Resiko Bencana/IRB) by 30%		5.1 Disaster Risk Index (Indeks Resiko Bencana/IRB) reaches 0% in all hamlets (Rukun Tetangga/RT)	Disaster Risk Index (Indeks Resiko Bencana/IRB) in all hamlets (Rukun Tetangga/RT)
			5.2 Increased community and institutional capacity in building urban resilience to climate change and disasters		5.2 Families receiving warnings in disaster-prone areas	Families receiving warnings in disaster prone areas
			5.3 Availability of early warning systems for weather and climate and disasters		5.3 Availability of early warning systems for weather and climate and disasters	Availability of early warning systems for weather and climate and disasters

(continued)

Table 5.1 (continued)

No	Global goal	Global target	National target	Village goal	Village target year 2030	Measurement
		6. By 2030, reduce the per capita environment-al impact of urban areas, including by paying special attention to air quality, including urban waste management	6.1 Increased coverage of urban waste management to 80%		6.1 Family waste management reaches 100%	Household waste management
			6.2 Realisation of a green city that is resilient to climate and disasters through the development and application of green water, green waste (waste management and waste through reduce-reuse-recycle), green transportation		6.2 There are waste treatment facilities in the village	Availability of waste treatment facilities in the village
		7. By 2030, provide public spaces and green open spaces that are safe, inclusive and easily accessible especially for women and children, the elderly and people with disabilities	7.1 Realisation of a green city that is resilient to climate, through provision of green open spaces		7.1 Existence of parks/ fields in the village	Existence of parks/ fields in the village
12	XII. Ensuring sustainable production and consumption patterns	1. Implement the 10-Year Framework of Programmes on Sustainable Consumption and Production Patterns, with all countries taking action, led by developed countries, considering the development and capacity of developing countries	1.1 The concept of a policy operational plan for sustainable consumption and production patterns and the development of its supporting system in the form of thematic collaboration quick wins program	12. Environment-ally Conscious Village Consumption and Production	1.1 The establishment of village regulations/ Head of Village Decree about business activities that do not cause pollution and management of waste and household waste	Village regulations/ Head of Village Decree about business activities that do not cause pollution and management of waste and household waste
		2. By 2020, achieve environment-ally friendly management of chemicals and all types of waste throughout their life cycle, in accordance with the agreed international framework and significantly reduce the pollution of these chemicals and waste to air, water, and soil to minimise adverse impacts on human health and the environment	2.1 Increased management of hazardous waste			
		3. By 2030, substantially reduce waste production through prevention, reduction, recycling, and reuse	3.1 Increased integrated waste management (reduce, reuse, and recycle/ 3R) through the operation of a city-scale recycling centre with a capacity of 20 tonnes per day		3.1 Availability of waste treatment units	Availability of waste treatment units

(continued)

Table 5.1 (continued)

No	Global goal	Global target	National target	Village goal	Village target year 2030	Measurement
		4. Encourage companies, especially large and transnational ones, to adopt sustainable practices and integrate sustainability information into reporting cycles	4.1 Increase in the number of companies implementing SNI ISO 14001 certification (Environmental Management System/EMS)			
		5. Enhancing sustainable public procurement practices, in accordance with national policies and priorities	5.1 Development of environmentally friendly products in the form of categories/criteria for products registered in public procurement (Green Public Procurement, GPP)			
13	XIII. Taking swift action to tackle climate change and its impacts	1. Strengthening resilience and adaptation capacity to climate-related hazards and natural disasters in all countries	1.1 Decrease in Disaster Risk Index through national and regional disaster risk reduction strategies	13. Climate Change Responsive Village	1.1 Disaster Risk Index reaches 0 in all hamlets (Rukun Tetangga/RT)	Disaster Risk Index in all hamlets (Rukun Tetangga/RT)
		2. Integrating climate change anticipatory actions into national policies, strategies and planning	2.1 Realisation of Greenhouse Gas (GHG) inventory management, as well as monitoring, reporting and verification of GHG emissions reported in the Biennial Update Report (BUR) document			
14	XIV. Conserve and sustainably use the oceans, seas and marine resources for sustainable development	1. By 2020, sustainably manage and protect marine and coastal ecosystems to avoid significant adverse impacts, including by strengthening their resilience, and take action for restoration to achieve healthy and productive oceans	1.1 Realisation of marine resource management governance through availability of policies, legal frameworks, and instruments related to national marine spatial planning, and sustainable management of fisheries management areas (FMA)	14. Village Caring for the Marine Environment	1.1 There is a village regulation/Head of Village Decree on marine resource protection There is a village regulation/Head of Village Decree on village spatial planning	Village regulation/Head of Village Decree on marine resource protection Village regulation/Head of Village Decree on village spatial planning
		2. In 2020, effectively regulate harvesting and stop overfishing, illegal fishing and destructive fishing practices, and implement science-based management plans to restore fish stocks as quickly as possible, at least to a level that can produce maximum sustainable yield according to its biological characteristics	2.1 Increase in production of capture fisheries		2.1 Increase in the volume of reasonable fish catches according to the type of fish	Volume of reasonable fish catches according to the type of fish

(continued)

Table 5.1 (continued)

No	Global goal	Global target	National target	Village goal	Village target year 2030	Measurement
		3. In 2020, preserve at least 10% of coastal and marine areas, consistent with national and international law and based on the best available scientific information	3.1 Increase in the area of water conservation areas		3.1 The area of water conservation areas is at least 33% of the village area	Area of water conservation areas compared to the village area
		4. In 2020, prohibit certain forms of fisheries subsidies that contribute to overcapacity and overfishing, eliminate subsidies that contribute to illegal, unreported and unregulated fishing and refrain from new types of subsidies, recognising that appropriate and effective special and differential treatment for developing and least developed countries should be an integral part of fisheries subsidies negotiations at the World Trade Organization	4.1 Control of Illegal, Unreported, Unregulated (IUU) fishing and damaging activities at sea marked by business actor compliance		4.1 No illegal fishing	Incidents of illegal fishing
15	XV. Protect, restore, and promote sustainable use of terrestrial ecosystems, sustainably manage forests, combat desertification, halt and reverse land degradation, and halt biodiversity loss	1. By 2020, ensure the conservation, restoration and sustainable use of terrestrial and freshwater ecosystems and their services, in particular forests, wetlands, mountains and drylands, in line with obligations under international agreements	1.1 Improve environmental quality through increased land/forest cover	15. Village Caring for the Land Environment	1.1 Area of open land planted is at least 33% of the village area	Area of open land planted compared to the area of the village
		2. In 2020, increase the implementation of sustainable management of all types of forests, halt deforestation, restore degraded forests and significantly increase afforestation and reforestation globally	2.1 Achievement of the area of degraded conservation areas restored to their ecosystem condition		2.1 The area of damaged forest land reaches 0% of the forest area	Area of damaged forest land compared to the forest area
			2.2 Increased efforts to utilise the results of forest wood restoration of the ecosystem		2.2 Utilisation of 100% of wood from the maintained/restored forest	Utilisation of wood from the maintained/ restored forest
		3. In 2020, halt desertification, restore land, and critical land, including land affected by desertification, drought and flooding, and strive to achieve a world free from degraded land	3.1 Reduction of critical land area through rehabilitation of 5.5 million hectares within the Forest Management Unit (Kesatuan Pengelolaan Hutan/KPH) and Priority Watershed (Daerah Aliran Sungai/DAS)		3.1 Critical land in the forest reaches 0% of the forest area	Critical land in the forest compared to the forest area

(continued)

Table 5.1 (continued)

No	Global goal	Global target	National target	Village goal	Village target year 2030	Measurement
		4. Take swift and significant action to reduce degradation of natural habitats, halt the loss of biodiversity, and by 2020, protect and prevent the extinction of threatened species	4.1 Increase of 10% in the population of endangered species		4.1 Increase of 50% in the population of endangered species	Percentage of the population of endangered species
		5. Take swift action to end hunting and trade of protected flora and fauna species and address demand and supply of wildlife products illegally	5.1 Increase in the percentage of completion of environmental crimes up to P21 (case ready to be submitted bu prosecutor to the court) by 5% per year from the number of cases that occur		5.1 Prosecuting environmental destroyers reaches 100%	Incidence of environmental destruction crimes
		6. In 2020, introduce measures to prevent the entry and significantly reduce the impact of invasive alien species on land and water ecosystems, and control or eradicate priority invasive alien species	6.1 Increase in quarantine and biological security systems			
		7. By 2020, integrate the values of ecosystems and biodiversity into national and regional planning, development processes, poverty reduction strategies and budgeting	7.1 Increase the utilisation of biodiversity to support economic growth, national competitiveness and community welfare		7.1 Availability of village regulations/ Head of Village Decree on biodiversity conservation	Village regulations/ Head of Village Decree on biodiversity conservation
16	XVI. Strengthen inclusive and peaceful societies for sustainable development, provide access to justice for all, and build effective, accountable, and inclusive institutions at all levels	1. Significantly reduce all forms of violence and death rates everywhere	1.1 Increase the sustainability efforts of social development characterised by controlled violence against children, fights, Domestic Violence (DV), and increased security reflected in small horizontal conflict and low crime rates	16. Village of Peace and Justice	1.1 Incidents of violence against children reach 0% Incidents of fights reach 0% Incidents of domestic violence (DV) reach 0% Incidents of crime reach 0%	Incidents of violence against children Incidents of fights Incidents of domestic violence (DV) Incidents of crime
					1.2. Mutual cooperation among residents of different religions, races, and groups	Mutual cooperation among residents of different religions, races, and groups

(continued)

Table 5.1 (continued)

No	Global goal	Global target	National target	Village goal	Village target year 2030	Measurement
		2. End abuse, exploitation, trafficking, and all forms of violence and torture against children	2.1 Decrease in the prevalence of violence against children		2.1 Incidents of violence against children reach 0%	Incidents of violence against children
					2.2 Child labour reaches 0%	Percentage of child labour in the village
					2.3 Human trafficking reaches 0%	Percentage of citizens traded in trafficking
		3. Promoting the rule of law at the national and international level and ensuring equal access to justice for all	3.1 Number of poor individuals or community groups who obtain legal aid for litigation and non-litigation		3.1 Number of poor individuals who obtain legal aid for litigation and non-litigation reaches 100%	Percentage of poor individuals who obtain legal aid for litigation and non-litigation
			3.2 Number of judicial services for the poor through court sessions outside the courthouse; waiver of case fees; and Legal Service Posts		3.2 Percentage of judicial services available for the poor through court sessions outside the courthouse reaches 100% Waiver of case fees applies to 100% of poor individuals Legal Service Posts are available	Percentage of judicial services for the poor through court sessions outside the courthouse Waiver of case fees applies to poor individuals Legal Service Posts
		4. Substantially reduce corruption and bribery in all its forms	4.1 Increase in the Anti-Corruption Behaviour Index		4.1 Anti-Corruption Behaviour Index reaches a score of 5	Anti-Corruption Behaviour Index

(continued)

Table 5.1 (continued)

No	Global goal	Global target	National target	Village goal	Village target year 2030	Measurement
		5. Develop effective, accountable, and transparent institutions at all levels	5.1 Increase in the percentage of Unqualified Opinion (Wajar Tanpa Pengecualian/WTP) on financial reports		5.1 The latest financial report is received in village deliberation forum Financial reports can be accessed by the public	Receipt of the latest financial report in village deliberation forum Access to financial reports for the public
			5.2 Increase in the percentage of B Score on the Government Agency Performance Accountability System (Sistem Akuntabilitas Kinerja Instansi Pemerintah/SAKIP)		5.2 The latest accountability report of the village head is received in village deliberation forum	Receiving the final accountability report of the village head during the village deliberation forum
			5.3 Increase in the use of E-procurement for procurement expenditure		5.3 The procurement process for goods and services for all development projects is open to the public	Transparency of the procurement process for goods and services for all village development projects
			5.4 Increase in the percentage of government agencies that have a good Bureaucratic Reform Index		5.4 Village government organisational structure in accordance with applicable laws and regulations	Compliance of the village government organisational structure with applicable laws and regulations
			5.5 Increase in the percentage of compliance with the Public Service Law		5.5 High level of citizen satisfaction with village government services	Level of citizen satisfaction with village government services

(continued)

Table 5.1 (continued)

No	Global goal	Global target	National target	Village goal	Village target year 2030	Measurement
		6. Ensuring responsive, inclusive, participatory and representative decision-making at all levels	6.1 Increase in the representation of women in the People's Representative Council and Regional People's Representative Council		6.1 Women in the Village Consultative Body (BPD) reach at least 30%	Percentage of women in the Village Consultative Body (BPD)
			6.2 Increase in the representation of women as decision-makers in executive institutions (Echelon I and II)		6.2 Percentage of women as village officials reaches at least 30%	Percentage of women as village officials
			6.3 Increase in the Democracy Institution Index		6.3 Democracy Institution Index reaches 100	Democracy Institution Index
			6.4 Increase in the Civil Liberties Index		6.4 Civil Liberties Index reaches 100	Civil Liberties Index
			6.5 Increase in the Political Rights Index		6.5 Political Rights Index reaches 100	Political Rights Index
		7. By 2030, provide legal identity for all, including birth registration	7.1 Increase in the coverage of basic birth certificate ownership services for the bottom 40% income population		7.1 Coverage of birth certificate ownership for the bottom 40% income population reaches 100%	Coverage of birth certificate ownership for the bottom 40% income population
			7.2 Increase in the percentage of children who have a birth certificate		7.2 Percentage of children who have a birth certificate reaches 100%	Percentage of children who have a birth certificate
		8. Ensure public access to information and protecting fundamental freedoms, in accordance with national regulations and international agreements	8.1 The realization of respect, protection, and fulfilment of human rights, which is demonstrated by the increasing handling of human rights violation complaints		8.1 Percentage of complaints of violations due to race, ethnicity, and religion that are handled reaches 100%	Percentage of complaints of violations due to race, ethnicity, and religion handled

(continued)

Table 5.1 (continued)

No	Global goal	Global target	National target	Village goal	Village target year 2030	Measurement
			8.2 The measurability of the Public Body in carrying out its obligations as regulated in Law No. 14 of 2008 on Public Information Disclosure, which is demonstrated by the increasing indicators of the obligation to announce public information, provide public information, manage and document public information, and public information		8.2 Development planning documents and financial report documents can be accessed by the public reaching 100% Summary of development planning documents and financial report documents are announced to the public in open spaces	Access to development planning documents and financial report documents Announcement of the summary of development planning documents and financial report documents in open spaces
			8.3 The implementation of the public information dispute resolution process through mediation and/or non-litigation adjudication with a percentage of 85% register per running year		8.3 Resolution of complaints related to public information reaches 100%	Percentage of resolution of complaints related to public information
			8.4 The increasing quality of Information and Documentation Management Officers (Pejabat Pengelola Informasi dan Dokumentasi/PPID) in carrying out their duties and functions as regulated in the legislation marked by the certification of PPID		8.4 The village government provides information on the same day	Length of time for the provision of village government information

(continued)

Table 5.1 (continued)

No	Global goal	Global target	National target	Village goal	Village target year 2030	Measurement
17	XVII. Strengthening the means of implementation and revitalising the global partnership for sustainable development	1. Strengthen the mobilisation of domestic resources, including through international support to developing countries, to increase local capacity for tax collection and other revenues	1.1 Achieving a tax revenue to GDP ratio above 12% per year	17. Partnership for Village Development	1.1 Achieving a village tax revenue to GDP ratio above 12% per year	Ratio of village tax revenue to GDP
		2. Enhance North–South, South– South and triangular regional and international cooperation and access to science, technology and innovation, and enhance sharing knowledge based on mutual agreement, including through better coordination between existing mechanisms, especially at the level of the United Nations (UN), and through global technology facilitation mechanisms	2.1 Increased implementation of South–South and Triangular development cooperation		2.1 There is inter-village cooperation There is village cooperation with the private sector There is cooperation with NGOs There is cooperation with international institutions	Inter-village cooperation Village cooperation with the private sector Cooperation with NGOs Cooperation with international institutions
			2.2 Availability of a national fibre optic backbone network that connects all major islands and districts/cities reaching 100% of districts/cities		2.2 Availability of internet network in the village Fast category internet speed	Availability of internet network in the village Village internet speed
			2.3 Increased penetration of fixed broadband access		2.3 There is wi-fi in the village Internet speed using the fast category	Availability of wi-fi in the village Internet speed using wi-fi
			2.4 Increased penetration of mobile broadband access		2.4 Internet usage with mobile phones reaches 100% Internet speed using the fast category	Internet usage with mobile phones Internet speed using mobile phones

(continued)

Table 5.1 (continued)

No	Global goal	Global target	National target	Village goal	Village target year 2030	Measurement
		3. Fully operationalising the technology and science bank, capacity building mechanism for technology and innovation for less developed countries by 2017 and increasing the use of enabling technology, especially information and communication technology	3.1 Availability of universal telecommunication access services and internet reaching 100% in the Universal Service Obligation (USO) area, with priority to remote, outermost, and border areas		3.1 Availability of internet network Internet speed using the fast category	Availability of internet network Internet speed
		4. Increase international support to implement effective and targeted capacity building in developing countries to support national plans to implement all sustainable development goals, including through North–South, South–South and triangular	4.1 Strengthening Indonesia's role in South–South and Triangular Cooperation (Kerjasama Selatan-Selatan dan Triangular/KSST)		4.1 There is village cooperation with international institutions	Village cooperation with international institutions
		5. Promoting a universal, rule-based, open, non-discriminatory and fair multilateral trading system under the World Trade Organization including through the conclusion of agreements under the Doha Development Agenda	5.1 Decrease in the average weighted tariff in Free Trade Agreement (FTA) partner countries (6 countries)			
		6. Significantly increase exports from developing countries, specifically with the aim of doubling the proportion of less developed countries in global exports by 2020	6.1 Increase in the growth of non-oil and gas product exports		6.1 Number of commodities from villages that are exported increase	The number of commodities from villages that are exported increases

(continued)

Table 5.1 (continued)

No	Global goal	Global target	National target	Village goal	Village target year 2030	Measurement
		7. Improve global macroeconomic stability, including through policy coordination and policy integration	7.1 Availability of a macroeconomic dashboard		7.1 Public access to information on village economic conditions	Access to information on village economic conditions
		8. Enhance global partnership for sustainable development, complemented by partnerships with various stakeholders that mobilise and share knowledge, expertise, technology and financial resources, to support the achievement of Sustainable Development Goals in all countries, especially in developing countries	8.1 Improving the quality of global cooperation to build mutual understanding between civilisations, and world peace, and to address global issues that threaten humanity		8.1 There is village cooperation with international institutions	Village cooperation with international institutions
		9. Encourage and enhance effective government–private and civil society cooperation, based on experience and sourced from cooperation strategies	9.1 Availability of alternative financing for development through the Public–Private Partnership (PPP)/ Government Cooperation with Business Entities in Infrastructure Provision (Kerjasama Pemerintah dan Badan Usaha/KPBU) scheme		9.1 There is village cooperation with the private sector	Village cooperation with the private sector
			9.2 Availability of state budget allocation for preparation, transactions and Government support for the PPP/PFI (Private Finance Intiative) project			
		10. In 2020, increase capacity development support for developing countries, including least developed and small island developing states, to significantly enhance the availability of high-quality, timely and reliable data, sorted by income, gender, age, race, ethnicity, migration status, disability, geographical location and other characteristics relevant to the national context	10.1 Increased availability of quality statistical data and information		10.1 Village statistical data available every year	Availability of village statistical data every year

(continued)

Table 5.1 (continued)

No	Global goal	Global target	National target	Village goal	Village target year 2030	Measurement
		11. By 2030, rely on existing initiatives to develop measurements of sustainable development progress that complement Gross Domestic Product, and support capacity development for statistics in developing countries	11.1 Increased competence of Human Resources (HR) in professional, integrity, and trustworthy statistics		11.1 Special officers handling village statistics are available	Availability of special officers handling village statistics
			11.2 Improved quality, quantity and use of Information and Communication Technology (ICT) facilities and infrastructure in statistical activities		11.2 Village statistical application available	Availability of village statistical application
			11.3 Increased availability and quality of accurate and timely development data and information, and the use of data and information for planning and evaluating development outcomes		11.3 Annual SDGs Desa data available Use of SDGs Desa data for Village Government Work Plan (Rencana Kerja Pemerintah Desa/ RKP Desa) and Village Revenue and Expenditure Budget (Anggaran Pendapatan dan Belanja Desa/APB Desa)	Availability of annual SDGs Desa data Use of SDGs Desa data for Village Government ork Plan and Village Revenue and Expenditure Budget
18				Dynamic Village Institutions and Adaptive Village Culture	1.1 Social piety	Percentage of mutual aid activities based on religious teachings
					1.2 Role of religious leaders in village deliberation forum and implementation of village development	Presence of religious leaders in village deliberation forum Participation of religious leaders in village development activities
					1.3 Village residents' concern for the care of orphans and the poor	Number of charity/ care activities for orphans and the poor
					2.1 The village government's organisational structure in accordance with the applicable legislation	The conformity of the village government's organizational structure with the applicable legislation

(continued)

Table 5.1 (continued)

No	Global goal	Global target	National target	Village goal	Village target year 2030	Measurement
					2.2 Implementation of village deliberation forum at least 4 times a year	Implementation of village deliberation forum
					2.3 Availability of Village Medium Term Development Plan	Availability of Village Medium Term Development Plan
					2.4. Availability of village boundary maps established by the regent/mayor	Availability of RKP Desa for the current year's implementation
					2.5 Availability of Village Work Plan for the current year's implementation	Availability of Village Work Plan for the current year's implementation
					2.6 Availability of Village Revenue and Expenditure Budget for the current year's implementation	Availability of Village Revenue and Expenditure Budget for the current year's implementation
					2.7 Accreditation level of Village-Owned Enterprises (Badan Usaha Milik Desa/ Bumdes)	Accreditation level of Village-Owned Enterprises (BAdan Usaha Milik Desa/ Bumdes)
					3.1. 100% of culture is preserved	Percentage of culture preserved
					3.2 Presence and participation of traditional institutions in village development	Presence of traditional institutions Participation of traditional institutions in village deliberation forum and implementation of village development

(continued)

Table 5.1 (continued)

No	Global goal	Global target	National target	Village goal	Village target year 2030	Measurement
					3.3 Percentage of social issues based on local culture resolved reaches more than 50%	Percentage of social issues according to local culture resolved
					4.1 The value of village assets increases every year	Type and value of village assets
					4.2 Percentage of village community institutions participating in village deliberation forum reaches more than 30%	Percentage of village community institutions participating in village deliberation forum

Chapter 6
SDGs Desa 1: Village Without Poverty

6.1 Construct of Village Without Poverty

Poverty remains a major challenge for the world's population, including Indonesia. Therefore, the main goal of development is to end poverty in all its forms everywhere. The World Bank defines poverty as a loss of well-being. This is characterised by a lack of key abilities, such as having insufficient income or education, poor health, feeling powerless, or lacking political freedom. The approach that can be taken is to ask whether households or individuals have sufficient resources to meet their needs. Do they have enough food? Or shelter? Or healthcare? Or education?

A Village without Poverty is a village where 100 percent of its inhabitants have the ability to meet their basic needs. The goal of this SDGs Desa is to target the poverty rate in the village to 0 percent by 2030. This means that by 2030 there will be no villagers classified as poor or below the Poverty Line.

In the RPJPN 2005–2025, the issue of poverty is viewed in a multidimensional framework; hence poverty is not only related to income measures, but also involves things such as: (i) the vulnerability and susceptibility of people or communities to become poor, (ii) the presence/absence of fulfilment of basic rights of citizens, and (iii) the presence/absence of differences in treatment of individuals or groups of people in living a dignified life.

According to data from the Indonesian Central Bureau of Statistics (Badan Pusat Statistik/BPS), the percentage of the population classified as poor in March 2023 was 9.36 percent, down 0.21 percentage points from September 2022 and 0.18 percentage points from March 2022. The number of poor people in March 2023 was 25.90 million, down 460 thousand from September 2022, and down 260 thousand from March 2022.

In regard to the area of residence of the poor, the percentage of the urban population classified as urban poor in March 2023 was 7.29 percent, down from 7.53 percent in September 2022. Meanwhile, the percentage the rural population classified as rural poor in March 2023 was 12.22 percent, down from 12.36 percent in September 2022.

A. H. Iskandar, *The Methods of Measuring the National Sustainable Development Goals in Indonesia*, Engaging Indonesia, https://doi.org/10.1007/978-981-96-0927-7_6

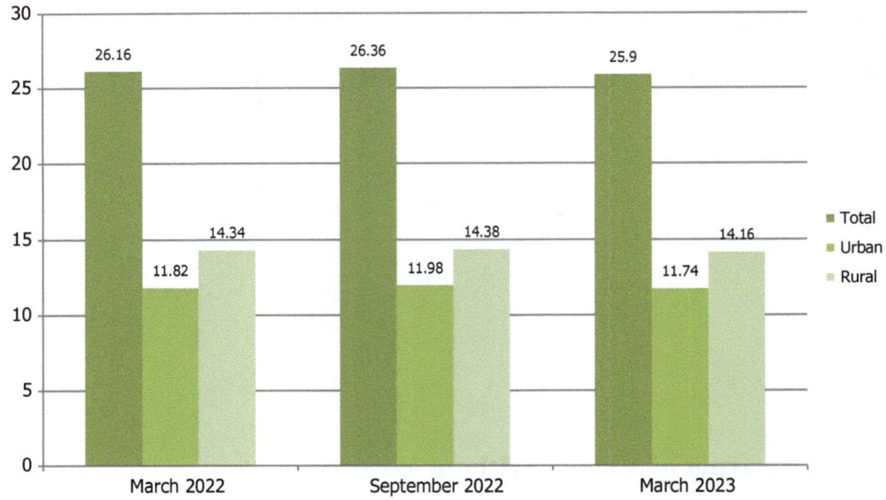

Fig. 6.1 Number of Poor People by Place of Residence, March 2022–March 2023 (in millions of people), *Source* BPS, 2023

Compared to March 2023, the number of urban poor in September 2023 decreased by 240 thousand people (from 11.98 million people in September 2022 to 11.74 million people in March 2023). Meanwhile, during the same period, the number of rural poor decreased by 220 thousand people (from 14.38 million people in September 2022 to 14.16 million people in March 2023). These numbers and percentages of urban and rural poor can be seen in Fig. 6.1.

The poverty line in March 2023 was recorded at IDR550,458 per capita per month with the composition of the Food Poverty Line being IDR408,522 (72.21 percent) and the Non-food Poverty Line being IDR141,936 (25.78 percent). Residents who have an average expenditure per capita per month below the poverty line are categorised as poor.

The Food Poverty Line (FPL) is the value of minimum food expenditure equivalent to 2,100 kilocalories per capita per day. The Non-food Poverty Line (NFPL) is the minimum requirement for housing, clothing, education and health.

In March 2023, the average poor household in Indonesia had 4.71 family members. Therefore, the average poverty line per poor household in Indonesia is IDR2,592,657 per poor household per month.

Based on this, the target of zero poor villagers is zero villagers below the Poverty Line. This means that the goal of a Village without Poverty is to alleviate 100 percent of villagers who are below the Poverty Line. This target must be achieved by 2030.

To achieve this target, many policies must be taken and implemented together, by the central government, regional government, and village government. These include increasing the income of the poor, ensuring access to basic services, and protecting the entire community from all forms of disaster.

6.2 Targets for Village Without Poverty

To achieve the goal of a Village without Poverty, 22 targets have been set that must be achieved by 2030. These targets are:

1. Poverty rate reaches 0%
2. The percentage of the population who become participants in health insurance through the National Social Security System (Sistem Jaminan Sosial Nasional/SJSN) for health sector reaches 100%
3. The number of poor disabled people receiving basic needs assistance increases year by year to 100%
4. The number of female-headed households (PEKKA) receiving basic needs assistance increases year by year to 100%
5. The number of poor families eligible for unconditional cash transfers increases year by year to 100%
6. The percentage of participation in the National Social Security System (SJSN) for employment sector reaches 100%
7. Coverage of childbirth in healthcare facilities for the lowest 40% income population reaches 100%
8. Coverage of complete basic immunisation for children aged 12–23 months for the bottom 40% of income-earning families reaches 100%
9. Coverage of contraceptive use in all ways for married productive-age women (aged 18–49 years) for the 40% of the lowest income population reaches 100%
10. Access to adequate drinking water for the lowest 40% income population reaches 100%
11. Access to adequate sanitation for the lowest 40% income population reaches 100%
12. The number of low-income households that can access decent housing reaches 100%
13. The Gross Enrolment Rate for primary school/equivalent in the village reaches 100%
14. The Gross Enrolment Rate for junior high school/equivalent in the village reaches 100%
15. The Gross Enrolment Rate for senior high school/vocational school/equivalent in the village reaches 100%
16. Ownership of birth certificates for the bottom 40% of income earners reaches 100%
17. Access to lighting for the bottom 40% of income earners reaches 100%
18. The number of locations for disaster risk reduction in hamlet/Neighbourhood Units (Rukun Tetangga/RT) reaches 100%
19. Basic needs for 100% of social disaster victims are met
20. Psychosocial support for 100% of social disaster victims
21. 100% of hamlets (Rukun Tetangga/RT) affected by natural/social disasters receive special services
22. Reduction of disasters in hamlet (Rukun Tetangga/RT).

6.3 Measurement and Analysis of a Village Without Poverty

In measuring the level of poverty on a macro scale, the Central Bureau of Statistics (BPS) uses the concept of the ability to meet basic needs (basic-needs approach). With this approach, poverty is viewed as an inability from an economic perspective to meet basic needs for food and non-food items, and is measured by expenditure. Based on this approach, the poor population is defined as the population with an average per capita expenditure per month below the Poverty Line.

The Poverty Line (PL) is the sum of the Food Poverty Line (FPL) and the Non-food Poverty Line (NFPL). The Food Poverty Line (FPL) is the expenditure value of minimum food needs equivalent to 2,100 kilocalories per capita per day. Among rural residents, the annual expenditure level for high food groups is:

1. Rice
2. Filtered clove cigarettes
3. Broiler chicken meat
4. Broiler chicken eggs
5. Granulated sugar
6. Tempeh
7. Tofu
8. Instant noodles
9. Shallots
10. Coffee
11. Bird's-eye chillies
12. Skipjack/tuna/mackerel
13. Bread
14. Wet cake
15. Beef.

The Non-food Poverty Line (NFPL) is the minimum requirement for housing, clothing, education and health. The annual expenditure for high non-food groups in rural areas is:

1. Housing
2. Electricity
3. Education
4. Transport
5. Petrol
6. Firewood
7. Ready-made children's clothes
8. Ready-made adult women's clothes
9. Bathing supplies
10. Health
11. Laundry soap.

In addition to using the Poverty Line as a macro measurement of poverty, poverty is also measured microscopically, referring to the assets owned by poor families. Micro poverty is usually used to directly target households in social assistance and social protection programmes. A community is considered poor if the household meets at least 9 out of the 14 criteria that follow (BPS):

1. The floor area of the dwelling is less than 8 m^2 per person
2. The type of dwelling floor is earth/bamboo/cheap wood
3. The type of dwelling wall bamboo/sago palm/low-quality wood/unplastered wall
4. Does not have a toilet facility/shares with other households
5. The household's source of lighting does not use electricity
6. The source of drinking water comes from a well/spring that is not protected/river/rainwater
7. The fuel for daily cooking is firewood/charcoal/kerosene
8. Only consumes meat/milk/chicken once a week
9. Only buys one set of new clothes a year
10. Only able to eat once/twice a day
11. Unable to afford medical expenses at healthcare centres/clinics
12. The head of the household's source of income is: a farmer with a land area of 500 m^2, farm labourer, fisherman, construction worker, plantation worker, or other job with an income below IDR 600,000 per month
13. The highest education level of the head of the household: did not go to school/did not complete primary school/completed primary school
14. Does not have savings/items that can be easily sold for a minimum of IDR 500,000, such as a motorbike on credit/non-credit, gold, livestock, motorboat, or other capital goods.

The synergy of measuring poverty on a macro and micro level will provide an overview of village strategies that can be implemented to achieve a village poverty level of 0%.

The measurements made by BPS can be used by villages to determine their own Poverty Line. The aspects of a Village without Poverty are presented in several formulas.

6.3.1 Poverty Level

Before measuring the poverty level, first income per capita per day should be measured. Income per capita per day is calculated from the income of all individuals in one year divided by the number of individuals, then divided again by 365 days. The formula is as follows:

$$\textit{Income per capita per day}(IDR) = \frac{\sum_{i=1}^{\textit{Number of Individuals}} \text{Income per Year}}{\sum \text{Individuals}} x \frac{1}{365}$$

The World Bank has set the standard for an extreme Poverty Line at 1.9 dollars PPP (Parity Purchasing Power), which indicates the number of rupiah spent to buy a certain amount of goods and services, based on price and quantity in each country. This is equivalent to IDR 11,000 per capita per day. This means, if the per capita income per day is below IDR 11,000, then that person is classified as extremely poor. Values below IDR 11,000 are included in the calculation of the extreme poverty rate.

The poverty rate, or the number of poor citizens, is obtained from the number of citizens with a per capita income per day that is below the BPS Poverty Line, divided by the number of individuals, multiplied by 100%. If the result is 0%, it can be said that the SDGs Village target has been achieved. The formula used is as follows:

$$\textit{Extreme Poverty Rate} = \frac{\sum_{i=1}^{\textit{Number of Individuals}} \textit{Individuals with income per capita per day} \leq \textit{IDR}12.000}{\sum \text{Individuals}} 100\%$$

$$\textit{Poverty Rate} = \frac{\sum_{i=1}^{\textit{Number of Individuals}} \textit{Individuals with income per capita per day} \leq \text{PL BPS}}{\sum \text{Individuals}} 100\%$$

The target poverty level of the village to be achieved is 0%. The village poverty level is measured based on the percentage of the population below the village Poverty Line. This target is measured using individual indicators, namely home ownership status (owned or rented), house structure (bamboo, wood, brick, soil, tile, cement, ceramic, etc.), as well as food expenditure in the last month and non-food expenditure in the last year. Therefore, actions that can be carried out in the village to achieve a village poverty rate of 0% include increasing the productivity and added value of commodities through reducing production input costs, and financial support to gain access to public services. For example, building village markets, village *online* shops, goods storage warehouses, and other things related to marketing support facilities. Other actions include developing village food reserves, whether it's rice, corn or sorghum through village barns.

6.3.2 National Social Security System for Health Sector

To measure the percentage of village residents participating in the Health Sector of the National Social Security System, the number of national health social security participants is divided by the total number of individuals with poor criteria, multiplied by 100%. If the result is 100%, then the SDGs Desa target is achieved. It can be calculated with the following formula:

$$\%\textit{National Health Social Security} = \frac{\sum \text{Participants in the National Health Social Security Programme with Poor Criteria}}{\sum \text{Individuals with Poor Criteria}} 100\%$$

The target percentage of the population who become social health insurance participants through the National Social Security System (SJSN) in the Health Sector is set at 100%. This target is measured using individual indicators, namely the number of families who are participants in the Social Security Agency on Health (Badan Penyelenggara Jaminan Sosial/BPJS) or other health insurance. To realize a village community with health insurance, all village residents must be encouraged to have BPJS or other health insurance. Activities that can be carried out are: socialisation and data collection regarding users or owners of Health Insurance Cards, and proposing Health Insurance Cards for village communities in order to realise 100% of village communities covered by health insurance.

6.3.3 Poor Disabled Recipients of Social Assistance

To measure the percentage of poor disabled recipients of social assistance, the number of poor disabled people who are involved in government programmes is divided by the total number of poor disabled people, then multiplied by 100%. If the result reaches 100%, then the SDGs Desa target has been achieved. The formula used is as follows:

$$\%\textit{Disabled poor recipients of social assistance} = \frac{\sum \text{Disabled poor recipients of social assistance}}{\sum \text{Poor disability}} 100\%$$

The target of poor disabled people who receive basic needs assistance, which increases from year to year, is set at 100%. This target is measured using individual indicators, namely the presence of disabled family members and the presence of disabled family members who receive government assistance, such as the Family Hope Programme (Program Keluarga Harapan/PKH), the Smart Indonesia Card (Kartu Indonesia Pintar/KIP), Social Assistance (Bantuan Sosial/Bansos), Unconditional Cash Transfers (Bantuan Langsung Tunai/BLT), among others. Increase of basic assistance to the disabled community can be achieved by: compiling proposals from disabled residents; carrying out socialisation and advocacy of facilities and infrastructure that are friendly to people with disabilities, and holding group discussions for disabled residents.

6.3.4 PEKKA Social Assistance Recipients

To measure the percentage of female-headed households (Perempuan Kepala Keluarga/PEKKA) that are recipients of social assistance, the number of PEKKA households that are involved in government programmes is divided by the total number of PEKKA households, then multiplied by 100%. If the result reaches 100%, then the SDGs Desa target has been achieved. The formula used is as follows:

$$\%PEKKA\ Social\ Assistance\ Recipients$$

$$= \frac{\sum \text{PEKKA Social Assistance Recipients}}{\sum \text{PEKKA}} 100\%$$

The target of PEKKA households receiving basic needs assistance has been increasing year by year, with a target of 100%. This target is almost the same as the previous one, which is aimed at marginal groups. This target is measured using individual indicators, namely the existence of PEKKA households and the existence of PEKKA households who receive government assistance, such as the Family Hope Program (Program Keluarga Harapan/PKH), Smart Indonesia Card (Kartu Indonesia Pintar/KIP), Social Assistance (Bantuan Sosial/Bansos), Unconditional Cash Transfers (Bantuan Langsung Tunai/BLT), and others. Increasing the portion of basic assistance given to PEKKA households can be done by: compiling proposals from PEKKA households who receive social assistance, conducting socialisation and advocacy of facilities and infrastructure that are friendly to women, and organising women's group discussions.

6.3.5 Poor Families Receiving Social Assistance

To measure the percentage of poor families receiving social assistance, the number of poor families receiving social assistance from government programmes is divided by the total number of poor families, then multiplied by 100%. If the result reaches 100%, then the SDGs Desa target has been achieved. The formula used is as follows:

$$\%Poor\ families\ receiving\ social\ assistance$$

$$= \frac{\sum \text{Poor families receiving social assistance}}{\sum \text{Poor families}} 100\%$$

The target of poor families eligible to receive unconditional cash transfers, which is increasing year by year, is set at 100%. This target is measured using household indicators, namely the presence of poor family members, as well as the presence of poor family members who receive social assistance from the government. Actions that the village government can carry out to improve the economic conditions of the community, enabling villagers to meet their daily needs even without Unconditional Cash Transfers (Bantuan Langsung Tunai/BLT), include: opening job vacancies in village business units, such as the Village-owned Enterprises (Badan Usaha Milik Desa/BUM Desa) work unit, and recruiting villagers for involvement in village infrastructure development activities, such as roads, bridges, and others, through the Cash-for-Work (Padat Karya Tunai Desa/PKTD) programme.

6.3.6 National Social Security System for Employment Sector

To measure the percentage of village residents participating in the SJSN for employment sector requires the number of employment social security participants with poor criteria and aged over 15 years to be divided by the total individuals with poor criteria aged over 15 years, then multiplied by 100%. If the result is 100%, then the SDGs Desa target is achieved. This measurement uses the following formula:

$$\%\mathit{National\ Employment\ Social\ Security}$$
$$= \frac{\sum \text{Participants in the national employment social security programme} > 15 \text{ yo with poor criteria}}{\sum \text{Individuals} > 15 \text{ yo with poor criteria}} 100\%$$

The target percentage of membership in the National Social Security System (SJSN) for employment sector is 100%. This target is measured using individual indicators, namely the number of family members who have employment insurance. Actions that can be carried out by the village government to encourage all villagers who work to have Social Security Agency on Health Card (Badan Penyelenggara Jaminan Sosial/BPJS) employment include, among others: carrying out data collection regarding people who do not yet have an Employment Guarantee Card; arranging socialisation on the use and ownership of employment guarantee cards; and organising collective proposal activities for the membership of the employment guarantee card for working villagers.

6.3.7 Coverage of Childbirth in Healthcare Facilities for the Lowest 40% Income Population

To measure the coverage of childbirth in health service facilities for the lowest 40% income population, the number of pregnant women who give birth in health service facilities with criteria of the lowest 40% income population divided by the number of pregnant women, then multiplied by 100%. If the result reaches 100%, then the SDGs Desa target is achieved. The formula used is as follows:

$$\% \ \textit{Childbirth in healthcare facilities}$$

For 40% of the lowest–income population

$$= \frac{\sum \begin{array}{c} \textit{Pregnant women criteria } 40\% \textit{ of the lowest-income population undergoing childbirth} \\ \textit{in healthcare facilities} \end{array}}{\sum \text{Pregnant women criteria } 40\% \text{of the lowest-income population}} 100\%$$

The target coverage of childbirth in healthcare facilities for 40% of the low-income population reaches 100%. This target is measured using an individual indicator, that is, the number of families in the bottom 40% who gave birth in the last year, the birthing facilities used (health centres/hospitals), and the birth assistants used (midwives/traditional birth attendants). Actionss that the village government can do to achieve this target include: improving facilities at village health and birthing centres; providing medical equipment and medicines for village health and birthing centres; and providing incentives to midwives and village health cadres to conduct health-check visits to low-income residents.

6.3.8 Full Basic Immunisation Coverage for Children Aged 12–23 Months for 40% of the Lowest-Income Families

To measure the full basic immunisation coverage for children aged 12–23 months for 40% of the lowest income families, the number of children aged 12–23 months who receive full basic immunisation with the criteria of 40% of the lowest income families divided by the number of children aged 12–23 months, then multiplied by 100%. If the result reaches 100%, then the SDGs Desa target is achieved. The formula used is as follows:

$$\% \ \textit{Basic complete immunisation for children aged } 12-23 \textit{ months}$$

For 40% of the lowest–income family

$$= \frac{\sum \begin{array}{c} \textit{Children aged } 12-23 \textit{ months receiving complete basic immunisation} \\ \textit{criteria } 40\% \textit{ of the lowest} - \textit{income family} \end{array}}{\sum \text{Children aged } 12-23 \text{ months criteria } 40\% \text{of the lowest} - \text{income family}} 100\%$$

The target for complete basic immunisation coverage in children aged 12–23 months for 40% of the lowest income families is set at 100%. This target is measured using individual indicators, namely the number of families who have children under 24 months, as well as the type of immunisation that has been given to the child in the last year. Actions carried out by the village government to support complete basic immunisation activities for children, among others, are: conducting socialisation and immunisation campaigns, conducting counselling activities, providing immunisation and Vitamin A through health workers, as well as conducting joint counselling activities with health workers related to child development, the role of parents in parenting, and others.

6.3.9 Coverage of Contraceptive Use in All Ways on Married Women of Productive Age (Aged 18–49 Years) for 40% of the Lowest Income Population

To measure the coverage of the usage rate of all methods of contraception among married women of productive age (18–49 years) for 40% of the lowest income population, the number of married women of productive age (18–49 years) who are active family planning participants (participants who are currently using one of the contraceptive tools) with the criteria of 40% of the lowest income population is required, divided by the number of married women of productive age (18–49 years), then multiplied by 100%. If the result reaches 100%, then the SDGs Desa target is achieved. The formula used is as follows:

$$\begin{aligned}&\%\ \textit{Contraceptive use by married women of productive age}\\&\textit{for } 40\% \textit{ of the lowest-income population}\\&= \frac{\sum \text{Married women of productive age}(18-49\text{yo})\text{who are active family planning participants criteria } 40\% \text{ of the lowest-income population}}{\sum \text{Married women of productive age}(18-49 \text{ yo}) \text{ criteria } 40\% \text{ of the lowest-income family}} 100\%\end{aligned}$$

The target coverage rate of contraceptive use by all methods, for married women of productive age (18–49 years), for the lowest 40% income population is set at 100%. This target is measured using individual indicators, namely the number of family members based on gender, age of family members, and married female family members who have used contraceptives in the past year. Actions that can be carried out by the village government, among others, are: campaign activities and socialisation about health, population, reproductive health and family planning, as well as conducting training for population cadres, family planning, and family development.

6.3.10 Access to Adequate Drinking Water for the Lowest 40% Income Population

To measure the percentage of people with access to adequate drinking water for the lowest 40% income population, the number of people who have access to adequate drinking water with the criteria of the lowest 40% income population is divided by the number of the lowest 40% income population, then multiplied by 100%. If the result is 100%, then the SDGs Desa target is achieved. The formula used is as follows:

$$\% \, Access \, to \, adequate \, drinking \, water = \frac{\sum \begin{array}{c} Population \, have \, gained \, access \, to \, adequate \, drinking \, water \\ criteria \, 40\% \, of \, the \, lowest-income \, population \end{array}}{\sum \text{Population criteria } 40\% of the lowest- \text{ income population}} 100\%$$

The target for access to potable water for the bottom 40% of the income population is set at 100%. This target is measured using individual indicators, namely the source of water used for drinking, bathing, and washing in the past year, as well as its quality (colour, taste, and smell). Actions that the village government can undertake include: building drinking water/clean water facilities managed by Village-owned Enterprise (BUM Desa) for the community, procuring pumps, constructing water storage tanks, building drinking water/clean water pipelines to residents' homes, procuring water tanks for households, collaborating with other villages in procuring water-tank trucks for distributing drinking/clean water to households not served by water pipelines, reforesting in water-source areas, and conducting drinking-water-management training.

6.3.11 Access to Adequate Sanitation for the Lowest 40% Income Population

To measure the percentage of adequate sanitation access for the lowest 40% income population, the number of people who have their own sanitation facilities with the criteria of 40% of the lowest income population divided by the number of 40% of the lowest income population, then multiplied by 100%. If the result is 100%, then the SDGs Desa target is achieved. The formula used is as follows

$$\% \text{Accesstoadequatesanitation} = \frac{\sum \begin{array}{c} \text{Population have their own facility for bathing, washing, and urinating/defecating} \\ \text{criteria 40\% of the lowest - income population} \end{array}}{\sum \text{Population criteria } 40\% of the lowest income population} 100\%$$

The target for access to decent sanitation for the lowest 40% income population is set to reach 100%. This target is measured using individual indicators, namely the type of defecation infrastructure, and wastewater drainage facilities for the lowest 40% income population. Actions carried out by the village government to improve facilities that support sanitation and encourage the community to use them, are, among others: building and rehabilitating public toilets for adults and children, building communal household sewage networks, conducting training activities for community health cadres related to sanitation, conducting campaign and socialisation activities related to sanitation, and conducting campaigns and socialisation in the community regarding how to use the toilets.

6.3.12 Low-Income Households that can Access Decent Housing

To measure low-income households that can access decent housing, the number of low-income households with decent housing conditions is divided by the number of low-income households, then multiplied by 100%. If the result is 100%, then the SDGs Desa target is achieved. The formula used is as follows:

$$\%\ Access\ to\ decent\ housing = \frac{\sum \text{Low} - \text{income households with decent housing condition}}{\sum \text{Low}-\text{income households}} 100\%$$

The target number of low-income households that can access decent housing is set at 100%. This target is measured using household indicators, namely home ownership status (owned or rented) and house structure (bamboo, wood, brick, soil, tile, cement, ceramic, etc.). Actions that can be carried out by the village government, among others, are: conducting data collection regarding low-income people whose houses are not decent, organising construction and/or home improvement activities for the poor, providing assistance with building materials for light, medium, and heavy rehabilitation, carrying out rehabilitation of houses not decent for low-income people using the Cash-for-Work (Padat Karya Tunai Desa/PKTD) approach, and conducting socialisation of the Decent House Programme assistance for low-income people from the government and local government.

6.3.13 Net Enrolment Rate of Primary School/Equivalent in the Village

To measure the ratio of the Net Enrolment Rate (NER) of primary school/equivalent, data on the number of children aged 7–12 years who are currently attending primary school/equivalent divided by the number of children aged 7–12 years, then multiplied by 100%. If the result is 100%, then the SDGs Desa target is achieved. The formula used is as follows:

$$\%\ NER\ of\ Primary\ School/Equivalent = \frac{\sum \text{Primary School/Equivalent Student aged } 7-12 \text{ years}}{\sum \text{Children aged } 7-12 \text{ years}} 100\%$$

6.3.14 Net Enrolment Rate of Junior High School/ Equivalent in the Village

To measure the Net Enrolment Rate (NER) of junior high school/equivalent, data on the number of teenagers aged 13–15 years who are currently attending junior high school/equivalent is required, divided by the number of teenagers aged 13–15 years, then multiplied by 100%. If the result is 100%, then the SDGs Desa target is achieved. The formula is as follows:

$$\% \ NER \ of \ Junior \ High \ Scool/Equivalent = \frac{\sum \text{Junior High}\textit{School/equivalent}\text{Student aged } 13-15 \text{ years}}{\sum \text{Teenager aged } 13-15 \text{ years}} 100\%$$

6.3.15 Gross Enrolment Rate of Senior High School/ Vocational High School/Islamic High School/ Equivalent in the Village

To measure the Gross Enrolment Rate of senior high school/vocational high school/ Islamic high school/equivalent, data on the number of teenagers aged 16–18 years who are currently attending high school/vocational high school/Islamic high school/ equivalent divided by the number of teenagers aged 16–18 years, then multiplied by 100%. If the result is 100%, then the SDGs Desa target is achieved. The formula is as follows:

$$\% \ GER \ of \ Senior \ High \ School/Equivalent = \frac{\sum \text{Senior High School/equivalent Student aged } 16-18 \text{ years}}{\sum \text{Teenager aged } 16-18 \text{ years}} 100\%$$

The target for the Gross Enrolment Rate of primary/elementary, junior high, and senior high school in the village is set to reach 100%. This target is measured using individual indicators, namely the birth dates of family members and the current educational levels of family members. Actions that can be undertaken to increase support for school facilities and to facilitate children who are unable to attend school include: providing school fee assistance for school-age children who are not enrolled in school or have dropped out; providing preparation equipment assistance for entering school for poor residents; providing educational cost assistance (transportation, money for buying books, uniforms) for children from poor families; providing educational cost assistance for children with special needs; and providing incentives for teachers.

6.3.16 Ownership of Birth Certificates for the Bottom 40% of the Income Population

To measure the ownership of birth certificates among the bottom 40% of the income population, the number of individuals within this group who have a birth certificate is divided by the total number of individuals within the bottom 40% of the income population, then multiplied by 100%. If the result is 100%, then the SDGs Desa target is achieved. The formula is as follows:

$$\%\ Ownership\ of\ Birth\ Certificate = \frac{\sum \text{The bottom } 40\% ofthepopulationbyincomewhohavebirthcertificates}{\sum \text{The bottom } 40\% ofthepopulation} 100\%$$

The target for birth certificate ownership among the bottom 40% income population is to reach 100%. This target is measured using individual indicators, specifically the number of family members who have birth certificates. Actions that the village government can undertake include: socialising about the importance of birth certificates, mobilising villagers to manage their birth certificates, and providing operational cost assistance for creating certificates for the underprivileged community, if needed.

6.3.17 Access to Lighting

To measure the percentage of access to lighting, the number of poor families who have already gained access to lighting is divided by the total number of poor families, then multiplied by 100%. If the result reaches 100%, then the SDGs Desa target has been achieved. The formula used is as follows:

$$\%Access\ to\ Lighting = \frac{\sum \text{Poor families have gained access to lighting}}{\sum \text{Poor families}} 100\%$$

The target for lighting access for the bottom 40% of the income population is set at 100%. This target is measured using the family indicator, namely the type of home lighting used (electric/non-electric), lighting-facility providers (State Electricity Company/PLN or private company), and use of lighting per day (in hours). Actions that can be undertaken include: building electricity lines for communities without electricity by utilising new renewable energy, such as the construction of micro hydropower plants (Pembangkit Listrik Tenaga Mikrohidro/PLTMH), diesel power, solar power, and wind power; building electrification networks managed by Village-owned Enterprises (Badan Usaha Milik Desa/BUM Desa); installing electrical systems in homes and providing subsidies for poor households, and conducting training in managing village electricity networks.

6.3.18 Location for Strengthening Disaster Risk Reduction in the Neighbourhood Units

To measure the location for strengthening disaster risk reduction in the hamlet/ Neighbourhood Units (Rukun Tetangga/RT), the number of hamlets (RT) with such locations is divided by the total number of hamlets (RT), then multiplied by 100%. If the result reaches 100%, the SDGs Desa target has been achieved. The formula used is as follows:

$$\% \, Location\ for\ strengthening\ disaster\ risk\ reduction\ in\ the\ HamletorNeighbourhood\ Units(RukunTetangga/R$$
$$= \frac{\sum \text{Location for strengthening disaster risk reduction in hamlet (RT)}}{\sum \text{Hamlet (RT)}} 100\%$$

The target for the number of disaster risk reduction reinforcement locations in the neighbourhood is set at 100%. This target is measured using the neighbourhood (Rukun Tetangga/RT) indicator, which considers the impact of natural disasters in the past year (floods, landslides, earthquakes, strong winds, tsunamis, and fires). Actions that can be carried out include: building disaster mitigation facilities to minimise the impact of disasters, creating announcement boards related to potential disasters and evacuation routes, building evacuation routes for disasters, such as floods, tsunamis, and volcanic eruptions, creating a map of potential disaster-prone areas in the village, and providing disaster information services.

6.3.19 Fulfilment of the Basic Needs of Social Disaster Victims

To measure the fulfilment of the basic needs of social disaster victims, the number of social disaster victims whose basic needs are met is divided by the number of social disaster victims, then multiplied by 100%. If the result reaches 100%, then the SDGs Desa target has been achieved. The formula used is as follows:

$$\% \, Fulfillment\ of\ the\ Basic\ Needs\ of\ Social\ Disaster\ Victims$$
$$= \frac{\sum \text{Fulfillment of the Basic Needs of Social Disaster Victims}}{\sum \text{Social Disaster Victims}} 100\%$$

The target of meeting basic needs of social disaster victims is set for 100%. This target is measured using individual indicators, as well as the availability of needs fulfilment during disasters, such as food, clothing, and shelter. Actions that can be carried out as emergency response actions when a disaster occurs are: building temporary shelters (refugee posts/shelters); building public kitchens; constructing clean water channels and emergency toilets; providing basic food aid during the emergency response period; providing standard first aid equipment (first aid box);

as well as providing special and urgent needs during disasters (for example, sanitary pads, baby diapers, baby milk, baby porridge, and so on) for groups of women, children, infants, toddlers, the elderly, the disabled and other vulnerable groups.

6.3.20 Psychosocial Support for Social Disaster Victims

To measure the psychosocial support of social disaster victims, the number of social disaster victims who receive psychosocial support is divided by the number of social disaster victims, then multiplied by 100%. If the result reaches 100%, then the SDGs Desa target has been achieved. The formula used is as follows:

$$\%\ Psychosocial\ support\ for\ social\ disaster\ victims$$
$$= \frac{\sum \text{Social disaster victims receiving psychosocial assistance}}{\sum \text{Social disaster victims}} 100\%$$

The target for psychosocial support is set at 100% for victims of social disasters. This target is measured using the individual indicator, which is the availability of disaster-trauma management from psychologists for affected families. Actions that can be carried out as efforts to heal fear/trauma after a disaster are: providing expert trauma healing and conducting trauma healing consultations for the community on a village scale.

6.3.21 Neighbourhood Units Affected by Natural Disasters/ Social Disasters Receive Special Services

To measure the neighbourhood unit/hamlets (Rukun Tetangga/RT) affected by natural disasters/social disasters receiving special services, the number of hamlets (Rukun Tetangga/RT) affected by natural disasters/social disasters receiving special services is divided by the number of hamlets (Rukun Tetangga/RT), then multiplied by 100%. If the result reaches 100%, then the SDGs Desa target has been achieved. The formula used is as follows:

$$\%\text{Hamlets (RT) affected by naturaldisasters or social disasters receive special services}$$
$$= \frac{\sum \text{Hamlets (RT) affected by natural disasters or social disasters receive special services}}{\sum \text{Hamlet (RT)}} 100\%$$

The target number of hamlets (Rukun Tetangga/RT) experiencing natural disasters/social disasters receiving special services is set at 100%. This target is measured using the Neighbourhood Units/hamlet (Rukun Tetangga/RT) indicator, namely the availability of disaster management from the government or private sector. Training activities that can be carried out related to potential disasters include: conducting community preparedness training in facing disasters; conducting volunteer training for disaster management; and conducting training on disaster potential recognition and disaster mitigation.

6.3.22 Reduction of Disasters in Neighbourhood Units

To measure the decrease in disasters in the hamlets (Rukun Tetangga/RT), data is needed on the number of hamlets (Rukun Tetangga/RT) that experienced natural disasters in a certain year, subtracting the number of RTs that experienced natural disasters in the previous year, dividing by the number of hamlets (Rukun Tetangga/RT) that experienced natural disasters in the previous year, then multiplying by 100%. If the result is negative (indicating a decrease), then the SDGs Desa target is achieved. The formula is as follows:

$$\%Reduction\ of\ disasters\ in\ hamlet(RT) = \frac{\sum \text{Natural disasters in a certain year} - \sum \text{Natural disasters in the previous year}}{\sum \text{Natural disasters in the previous year}} 100\%$$

The target for disaster occurrences in hamlets (Rukun Tetangga/RT) is set to decrease. This target is measured using the hamlets (Rukun Tetangga/RT) indicator, which is the number of disasters in the last year (floods, landslides, earthquakes, strong winds, tsunamis, and fires). This target is also measured using the head of household/individual indicator, namely the impact of natural disasters in the past year. Activities that can be carried out include creating/improving ditches in flood-prone areas, providing flood control equipment, creating breakwater embankments, and greening activities on landslide-prone mountain slopes, or with concrete landslide retaining structures (retaining walls).

6.4 Operational Definitions for Village Without Poverty

Key words used related to the achievement of SDGs Desa 1, Village without Poverty, are shown in Table 6.1.

Table 6.1 Operational definitions for village without poverty

	Key word	Operational definition
1.	Owned Assets	Wealth owned that has exchange value
2.	Cooking Fuel	Fuel used for cooking, such as: kerosene, gas, electricity, coal, briquettes, charcoal, firewood, and others
3.	Welfare Rice Assistance	Non-cash assistance in the form of foodstuffs (rice) provided by the government to poor families every month
4.	Social Security Agency of Health (BPJS Kesehatan)/Smart Indonesia Card (Kartu Indonesia Pintar/KIP) Contribution Assistance	Government assistance in the form of contributions so that poor citizens get social health insurance/government assistance in the form of a card that provides guarantees for poor families to get health service benefits
5.	Family Hope Programme Assistance (PKH)	Conditional social assistance to poor families determined as beneficiary families that provide access for poor families, especially pregnant women and children, to utilise various health service facilities (faskes) and education service facilities (fasdik) available around them
6.	Social disaster	Social disaster is a disaster caused by an event or series of events caused by humans that include social conflicts between groups or communities, and terror
7.	Unconditional-cash-transfer (Bantuan Langsung Tunai/BLT)/ temporary-unconditional-cash-transfer for community (Bantuan Langsung Sementara Masyarakat/BLSM)	Direct Cash Assistance/Temporary Direct Assistance for the community. A government assistance programme in the form of cash grants, either conditional (conditional cash transfer) or unconditional (unconditional cash transfer), temporary in nature to ensure the community can still meet their daily needs
8.	Corn	A carbohydrate-producing plant (Latin name: Zea mays ssp) with a flower arrangement that has male flowers arranged in the form of tassels and female flowers in the form of cobs
9.	Number of family members	The number of family members who are dependents living in one house
10.	Family	The smallest unit in society consisting of a husband and wife, or husband, wife and their children, or father and his children, or mother and her children (Law No 52 of 2009)

(continued)

Table 6.1 (continued)

	Key word	Operational definition
11.	Head of Family	A man or woman who is married, or a widow/widower who heads or is responsible for a family whose members consist of his/her spouse and or their children
12.	Family has a clean drinking water source	A family that has a source of drinking water consisting of bottled water, refill water, piped water, and well water/pump and protected sources and protected springs
13.	Family has its own toilet facilities	Availability of a latrine/toilet for each family
14.	Monthly income (IDR)	Receipt of money for activities carried out or services provided in one month in rupiah units
15.	Cash-for-Work programme	A programme in the form of community empowerment activities in villages, especially for the poor and marginalised, which are productive in nature and prioritise the use of resources, labour, and local technology to increase income, reduce poverty, and improve people's welfare (according to related regulations)
16.	Source of Lighting	The source of lighting used, such as candles, lanterns, oil lamps, electric lamps, kerosene lamps, and others

Chapter 7
SDGs Desa 2: Village Without Hunger

7.1 Concept of a Village Without Hunger

Village without Hunger aims to eliminate hunger among village residents by 2030. This goal will be achieved through several objectives, such as establishing food sovereignty, improving nutrition, and promoting sustainable agriculture.

Hunger is an uncomfortable or painful physical sensation caused by insufficient food energy consumption (FAO 2021). This condition becomes chronic when it shifts to food insecurity, that is, when a person does not consume enough food energy calories regularly, to lead a normal, active, and healthy life. Food insecurity has several levels of severity measured by the Food Insecurity Experience Scale (FIES).

According to the Global Hunger Index (2020), the level of hunger in Indonesia continues to decline. In 2020, Indonesia's hunger level was 19.1, down 1 point from 20.1 in 2019. 2020 was the first year Indonesia moved out of the 'Serious' category and into the 'Moderate' category (range 20–34.9). The decline in the level of hunger in Indonesia, is shown in Fig. 7.1.

The decrease in the level of hunger is influenced, among other things, by the decrease in the global poverty rate from 28.6 percent in 1999 to 9.9 percent in 2019. However, compared with ASEAN countries, hunger in Indonesia is still worse than in Thailand, Malaysia, Vietnam, and the Philippines (Lokadata, 2020). The Central Bureau of Statistics (2019) recorda that the malnutrition rate has reached 30,000, or, in other words, one person out of every 10,000 residents suffers from malnutrition.

Factually, in daily life, not everyone has the same opportunity to get the food they need. Therefore, the key to eradicating hunger, food insecurity, and all forms of malnutrition is to improve access to food and increase sustainable agricultural production. This includes increasing farmer productivity and income; developing technology and market access; establishing sustainable food production systems; and creating added value for agricultural production.

In short, the goal of Village without Hunger is to achieve a village condition with 0 percent of village residents experiencing physical discomfort caused by a lack of food energy consumption. This means that by 2030, the goal of this SDGs

A. H. Iskandar, *The Methods of Measuring the National Sustainable Development Goals in Indonesia*, Engaging Indonesia, https://doi.org/10.1007/978-981-96-0927-7_7

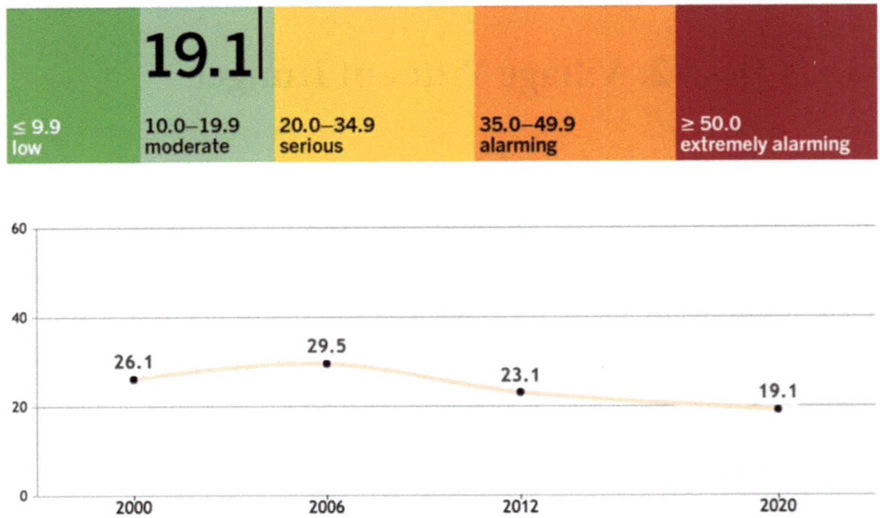

Fig. 7.1 Score GHI Indonesia, 2000–2020

Desa is to ensure that 100 percent of village residents have sufficient food energy consumption, the village is not in a state of food insecurity, and there are no village residents suffering from malnutrition.

7.2 Village Without Hunger Targets

To achieve the goal of a Village without Hunger, eight targets have been set to be achieved by 2030. These targets are:

(1) Prevalence of undernutrition (underweight) in children under five years old (toddlers) reaches 0%.
(2) The proportion of the population with a minimum calorie intake below 1,400 kcal/capita/day reaches 0%.
(3) The prevalence of stunting (short and very short) in children under two years old (infants) reaches 0%.
(4) The prevalence of wasting (thinness) in toddlers reaches 0%.
(5) The prevalence of anaemia in pregnant women reaches 0%.
(6) Infants under the age of 6 months receiving exclusive breastfeeding reaches 100%.
(7) The quality of food consumption indicated by the Food Pattern Hope (Pola Pangan Harapan/PPH) score reaches 100%.
(8) There are sustainable food farming areas.

7.3 Measurement and Analysis of Village Without Hunger

To see the extent of the hunger aspect in the village, the village can measure using the targets of Village without Hunger. The data required are children with malnutrition, residents with minimum calorie intake, sufferers of diseases due to hunger, and other data related to the aspect of hunger in the village. Data on food supply, such as the existence of food farming areas in the village, is also needed. To measure the achievement of the targets of Village without Hunger, several formulas are presented as follows:

7.3.1 Prevalence of Malnutrition

The measurement of the prevalence of malnutrition in children under five years (toddlers) can be done by the village itself. The prevalence of toddlers who experience malnutrition is measured by adding up all toddlers who experience malnutrition divided by the number of toddlers, multiplied by 100%. If the result is 0%, then one of the SDGs Desa targets regarding Village without Hunger, has been achieved. The following formula can be used to measure the prevalence of malnutrition in toddlers in the village.

$$\%\ Undernourished\ Toddlers = \frac{\sum \text{Undernourished toddlers}}{\sum \text{Toddlers}} \times 100\ \%$$

The target prevalence of undernutrition (underweight) in infants and children is targeted to reach 0%. This target is measured using individual indicators, namely the number of children under the age of one year, three years, and five years who have suffered from malnutrition in the past year, as well as the number of children under the age of one year, three years, and five years in the past year. Actions to improve community nutrition especially for infants, toddlers, and children are: providing additional healthy food for improving nutrition in toddlers; conducting socialisation related to children's nutritional needs; and providing financial assistance for care of toddlers suffering from severe malnutrition.

7.3.2 Population with Minimum Calorie Intake Below 1,400 kcal/Capita/Day

The proportion of the population with a minimum calorie intake below 1,400 kcal/capita/day is measured by summing up the population with a minimum calorie intake below 1,400 kcal/capita/day in the village, divided by the number of people in the village, then multiplied by 100%. If the result is 0%, then one of the SDGs Desa targets regarding Village without Hunger, is achieved. The following formula can be used to measure the proportion of the population with a minimum calorie intake below 1400 kcal/capita/day.

$$\% \textit{ Population with Minimum Calorie Intake} = \frac{\sum \text{Population with minimum calorie intake below 1400 kcal/capita/day}}{\sum \text{Population}} \times 100\%$$

The target proportion of the population with a minimum calorie intake is targeted below 1,400 kcal/capita/day. This target is measured using an individual indicator, namely the number of meals 1–2 times a day for the last month. Actions by the village government to improve nutrition in low-income communities are: providing nutritious supplementary food and food assistance for low-income families; optimising vacant land around residents' homes to plant crops; using village funds for the procurement of seeds, agricultural production facilities; among others.

7.3.3 Prevalence of Stunting

The nutritional status of children, including stunting, can be assessed at healthcarefacilities, such as Community Health Centres (Puskesmas) and Integrated Service Posts (Posyandu). Village residents can also calculate the length/height of the child and then compare it with the Child Anthropometry Standards set by the Ministry of Health. Furthermore, to measure the prevalence of stunting, the number of toddlers suffering from stunting is divided by the total number of toddlers, then multiplied by 100%. If the result is 0%, then one of the SDGs Desa targets regarding Village without Hunger is achieved. The following formula can be used to measure the prevalence of stunting.

$$\textit{Stunting prevalence} = \frac{\sum \text{Toddlers with Stunting}}{\sum \text{Toddlers}} \times 100\%$$

The target prevalence of stunting (short and very short) in infants and children is targeted at 0%. This target is measured using individual indicators, namely the height of children under the age of one year, three years, and five years. Actions to reduce the chances of children being born with stunting and to treat children with stunting are: providing financial assistance for infants and children suffering from stunting from low-income families and providing additional healthy food according to treatment recommendations. For the prevention of stunting in toddlers, pregnant and breastfeeding women are provided with consultation services and free examinations for toddlers so they can work together with health centres or local health workers who conduct socialisation, and carry out a stunting-free campaign that includes advice on both prevention and treatment.

7.3.4 Prevalence of Wasting

The assessment of a child's nutritional status, including wasting, can be carried out at healthcare facilities, for example Community Health Centres and Integrated Service Posts. Village residents can also calculate the weight of the child then compare it with the Child Anthropometry Standards set by the Ministry of Health. Furthermore, to measure the prevalence of wasting, the number of toddlers who suffer from wasting is divided by the total number of toddlers, then multiplied by 100%. If the result is 0%, then one of the SDGs Desa targets regarding Village without Hunger is achieved. The following formula can be used to measure the prevalence of stunting:

$$Wasting\ Prevalence = \frac{\sum \text{Toddlers with Wasting}}{\sum \text{Toddlers}} \times 100\%$$

The target prevalence of wasting (thinness) in infants and children is aimed to reach 0%. This target is measured using individual indicators, namely the weight of children under the age of one year, three years, and five years. Actions by the village government to improve nutrition in the community, especially for infants, toddlers, and children, among others, are: providing additional healthy food to improve nutrition for toddlers; conducting socialisation related to children's nutritional needs; and providing financial assistance for the care of toddlers suffering from malnutrition.

7.3.5 Prevalence of Anaemia in Pregnant Women

To measure the prevalence of anaemia in pregnant women, the number of pregnant women suffering from anaemia is divided by the total number of pregnant women, then multiplied by 100%. If the result is 0%, then one of the SDGs Desa targets regarding Village Without Hunger is achieved. The following formula can be used to measuring the prevalence of anaemia in pregnant women.

$$Prevalence\ of\ anaemia\ in\ pregnant\ women = \frac{\sum \text{anaemic pregnant women}}{\sum \text{pregnant women}} \times 100\%$$

The target prevalence of anaemia in pregnant women is aimed to reach 0%. This target is measured using individual indicators, namely the number of pregnant women suffering from anaemia. Actions to provide health assistance/support to pregnant women include providing free vitamin service assistance and additional healthy food for pregnant women from low-income families who are suffering from anaemia.

7.3.6 *Infants Under Six Months Receiving Exclusive Breastfeeding*

The percentage of infants under six months receiving exclusive breastfeeding is measured by adding up infants under six months who receive exclusive breastfeeding, divided by the number of infants in the village, then multiplied by 100%. If the result is 100%, then one of the SDGs Desa targets regarding Village without Hunger is achieved. The following formula can be used for the percentage of infants under six months receiving exclusive breastfeeding in the village:

$$\%\ \textit{Infants receiving exclusive breastfeeding} = \frac{\sum \text{Infants receiving exclusive breastfeeding}}{\sum \text{Infants}} \times 100\%$$

The target for babies under six months to receive exclusive breastfeeding is 100%. This target is measured using individual indicators, namely the number of babies receiving exclusive breastfeeding (aged from one to six months). Actions to provide health support to breastfeeding mothers include, among others, providing additional healthy food for breastfeeding mothers from low-income families, as well as providing free consultation services for breastfeeding mothers related to exclusive breastfeeding.

7.3.7 *Quality of Food Consumption Indicated by the Score of the Desired Dietary Pattern (DDP)*

In 1989, UNICEF's Food and Agriculture Organisation (FAO) introduced the Desirable Dietary Pattern (DDP) or Desired Dietary Pattern (DDP), which is the composition of the main food groups that, when regularly consumed, can meet energy needs and provide other important nutrients. The DDP can be used as a tool to assess the food situation of the population expressed in DDP scores. The higher the DDP score (maximum 100), the more diverse and balanced the food consumption. The DDP score is used as an indicator of nutritional quality and diversity of food consumption so it can be used to plan food consumption needs in the coming years. The DDP can be used as a guideline in the evaluation and planning of the food supply, and the food production and consumption of the population, both in terms of quantity, quality, and diversity by considering social, economic, cultural, religious and taste aspects (Ministry of Agriculture, 2015). If the DDP score is 100, then one of the SDGs Desa targets regarding Village without Hunger is achieved. Table 7.1 shows the composition of the National DDP.

The target for food consumption quality, indicated by the Food Pattern Hope (Pola Pangan Harapan/PPH) score, is set to reach 100%. The success of this target is measured using family/individual indicators, namely the consumption of rice and fish

Table 7.1 Composition of the national DDP

No	Food group	% RDA (FAO RAPA)	National desired dietary pattern				
			Gram	Energy (kcal)	% RDA	Weight	PPH Score
1	Cereals	40–60	275	1075.0	50.0	0.5	**25.0**
2	Tubers	0–8	100	129.0	6.0	0.5	**2.5**
3	Animal Food	5–20	150	258.0	12.0	2.0	**24.0**
4	Oils and Fats	5–15	20	215.0	10.0	0.5	**5.0**
5	Oily Fruits/ Seeds	0–3	10	64.5	3.0	0.5	**1.0**
6	Legumes	2–10	35	107.5	5.0	2.0	**10.0**
7	Sugar	2–15	30	107.5	5.0	0.5	**2.5**
8	Vegetables and Fruits	3–8	250	129.0	6.0	5.0	**30.0**
9	Others	0–5	–	64.5	3.0	0.0	**0.0**
Total				**2,150**	**100.0**	–	**100.0**

Source Ministry of Agriculture (2015)

per person per day, or expenditure for eating in the last month. To achieve this target, village governments can carry out actions including campaigns for the community promoting good food patterns, balanced nutrition, and fish consumption.

7.3.8 *Existence of Food Agriculture in Each Neighbourhood Units*

To measure the existence of sustainable food agricultural areas, the number of irrigated rice fields in the Neighbourhood Units/hamlets (Rukun Tetangga/RT) plus the number of non-irrigated rice fields in the hampets (Rukun Tetangga/RT) is required, the sum of which is divided by the number of hamlets (Rukun Tetangga/RT) in the village, then multiplied by 100%. If the result is above 0%, then one of the SDGs Desa targets regarding Village without Hunger is achieved. The formula that can be used is as follows:

$$\%\,\text{Existence of Food Agriculture in Each Neighbourhood Units} = \frac{\sum \text{Irrigated and non-irrigated rice fields per hamlet (RT)}}{\sum \text{hamlet (RT)}} \times 100\%$$

The final target in the Village without Hunger is the existence of a sustainable food agriculture area. This target is measured by the availability of sustainable food agriculture areas in the village and the RTs. To achieve this target, actions can be carried out by the village government including establishing a sustainable agriculture area.

7.4 Operational Definitions for Village Without Hunger

Key words related to the Village without Hunger, SDGs Desa 2, are found in Table 7.2.

Table 7.2 Operational definitions for village without hunger

	Keyword	Operational definition
1	Anaemia	A condition when the body lacks healthy red blood cells or when red blood cells do not function properly. As a result, the body's organs do not get enough oxygen, making the anaemia sufferer pale and easily tired
2	Under Two Years (Bawah Dua Tahun/Baduta)	Children under two years of age
3	Under Five Years (Bawah Lima Tahun/Balita)	Children aged one to five years
4	Under Three Years (Bawah Tiga Tahun/ Batita)	Children under three years of age
5	Infant	An individual aged 0–12 months characterised by rapid growth and development accompanied by changes in nutritional needs
6	Nutrition	Organic substances required by organisms for the normal function of the body system, growth, health maintenance
7	Sustainable Food Agriculture Area	An agricultural cultivation area mainly in rural regions that has a stretch of Sustainable Food Agriculture Land and/or a stretch of Reserve Food Agriculture Land and its supporting elements with the main function to support self-reliance, resilience, and national food sovereignty
8	Monthly expenditure (IDR)	Money given for goods or services obtained in one month in rupiah units
9	Paddy field	Agricultural land that is divided into plots and bounded by embankments or irrigation channels, usually planted with paddy, secondary crops, or other cultivated plants
10	Irrigated paddy field	Paddy field that has an irrigation network where the supply channel is separate from the drainage channel so that the provision and distribution of water in the paddy field can be fully regulated and easily measured
11	Child Anthropometry Standards	A collection of data about size, proportions, and body composition as a reference for assessing nutritional status and growth trends of children
12	Stunting (short toddler)	Characterised by a lack of height/length for age (H/A). Body length is used for children under 24 months and height for children aged 24 months and above. Short toddlers are caused by long-term conditions, so a characteristic shown by short toddlers is chronic nutritional problems
13	Wasting (thin toddler)	Characterised by a lack of weight for height/length (W/H). Body length is used for children under 24 months and height for children aged 24 months and above. Thin toddlers are caused by a lack of food or by infectious diseases that occur over a short period. A characteristic shown by thin toddlers is acute nutritional problems

References

FAO (2021) The state of food insecurity and nutrition in the world. Available at: https://www.fao.org/3/cb4474en/cb4474en.pdf

Global Hunger Index (2021) Indonesia. Available at: https://www.globalhungerindex.org/indonesia.html

Ministry of Agriculture (2015) Guide to calculating the food Pattern Hope (PPH). Available at: http://bkp.pertanian.go.id/storage/app/media/Evalap/BUKU%20PEDOMAN%20PENYUSUNAN%20PPH.pdf

Chapter 8
SDGs Desa 3: Healthy and Prosperous Village

8.1 Concept of Healthy and Prosperous Village

After ending Village without Poverty and Village without Hunger, the next SDGs Desa goal is Healthy and Prosperous Village. This goal ensures the realisation of village conditions that guarantee the healthy and prosperous life of village residents. Therefore, various efforts to achieve this goal are focused on the availability of easy access to health services for village residents.

The World Health Organisation defines health as a state of complete physical, mental, and social well-being, not just as freedom from disease or physical and mental weakness. Meanwhile, prosperity according to the Great Dictionary of the Indonesian Language, means feeling peaceful, safe and sound, and happy. The concepts of health and prosperity cannot be separated. A village with a healthy community is expected to live prosperously.

Data from the Ministry of Health (2018) presents household opinions about easy access to hospitals, regarding the type of transportation, travel time, and cost, and shows that 14.6% of residents find it easy, 42.4% find it difficult, and 43% find it very difficult. Furthermore, easy access to Community Health Centres/Village Health Posts/Mobile Health Centres/Village Midwives is easy for 31% of residents, difficult for 32.2%, and very difficult for 36.8%.

Therefore, to realise a Healthy and Prosperous Village, the village and supra-village governments must ensure easy access for villagers to health services, affordable health insurance for villagers, a decrease in maternal mortality rates (MMR), a decrease in infant mortality rates (IMR), an increase in the provision of complete immunisation for infants, an increase in the prevalence of contraceptive use, control of HIV/AIDS, TB, obesity, malaria, leprosy, filariasis (elephantiasis), control of drug abuse, and a decrease in the birth rate for teenage mothers.

The Healthy Indonesia Programme is one of the government's health sector programmes. This programme has three main pillars, set out by the Ministry of Health (2017) as follows:

A. H. Iskandar, *The Methods of Measuring the National Sustainable Development Goals in Indonesia*, Engaging Indonesia, https://doi.org/10.1007/978-981-96-0927-7_8

(1) The application of a health paradigm is carried out with a strategy of mainstreaming health in development, strengthening promotive and preventive efforts, and empowering the community.
(2) Strengthening health services is carried out and directed towards improving access to and quality of services. In terms of primary health services, it is directed towards promotive and preventive service efforts, through a continuum of care approach and risk-based health interventions in clinical governance, management governance, and programme governance.
(3) The implementation of National Health Insurance (NHI), as the state is determined to guarantee health services for all residents and foreign nationals living in Indonesia.

8.2 Healthy and Prosperous Village Targets

To achieve the goal of a Healthy and Prosperous Village, the following 23 targets have been set for 2030:

1. Coverage of National Health Insurance (NHI) reaches 100%
2. Unmet need for health services reaches 0%
3. The percentage of childbirths in health facilities reaches 100%
4. The percentage of childbirths by skilled health workers reaches 100%
5. Maternal Mortality Rate per 100,000 live births reaches 0%
6. Infant Mortality Rate per 1,000 live births reaches 0%
7. The percentage of households with complete basic immunisation for infants reaches 100%
8. HIV prevalence in the adult population reaches 0%
9. Tuberculosis (TB) prevalence per 100,000 population reaches 0%
10. Prevalence of high blood pressure reaches 0%
11. Prevalence of obesity in the population aged 18 and over reaches 0%
12. Prevalence of drug abuse reaches 0%
13. The number of households with malaria elimination reaches 100%
14. The number of households with leprosy elimination reaches 100%
15. The number of households with filariasis (elephantiasis) elimination reaches 100%
16. The percentage of smoking in the population aged $\leq$ 18 years reaches 0%
17. Increase in the number of households that have access to a community health centre that supports mental health
18. The number of victims of drug abuse who receive social rehabilitation in shelters according to service standards reaches 100%
19. The number of victims of death and serious injuries due to traffic accidents reaches 0%
20. The prevalence of short-term contraceptive use in married people of reproductive age (ages 18–49 years) reaches 100%

21. The rate of use of long-term contraceptive methods (LARC) in a modern way in married people of reproductive age (ages 18–49 years) reaches 100%
22. The birth rate in adolescents aged 15–19 years (Age Specific Fertility Rate/ ASFR) reaches 0%
23. Total Fertility Rate (TFR) below 1.5.

8.3 Measurement and Analysis of Healthy and Prosperous Villages

Health is one of the variables for improving the quality of human resources. Therefore, to improve the quality of human resources in the village, one of the things that must be considered is the health of the villagers. Through the measurement of Healthy and Prosperous Village targets, the village can develop programmes and activities to improve the quality of village human resources, especially related to health, and can evaluate the health of the village community alongside the availability of health facilities in the village. To measure the achievement of Healthy and Prosperous Village targets, several formulas are presented as follows.

8.3.1 National Health Insurance

To measure the percentage of villagers participating in National Health Insurance (Jaminan Kesehatan Nasional/JKN), the number of villagers participating in JKN is divided by the number of village residents, then multiplied by 100%. If the result is 100%, then the SDGs Desa target is achieved. Thus, the extent of JKN coverage can be calculated with the following formula:

$$\% \ National\ Health\ Insurance\ (JKN) = \frac{\sum \text{Village residents participatin in JKN}}{\sum \text{Village residents}} \times 100\%$$

The target coverage of the National Health Insurance (JKN) is set to reach 100% by 2030. The indicators used are the number of National Health Insurance (JKN) holders in the last year, as well as the number of National Health Insurance (JKN) holders in the previous year. Meanwhile, activities that can be carried out to encourage all residents to have JKN are organising socialisation of the use and ownership of the National Health Insurance (JKN) card, as well as advocating for the proposal of health insurance cards for village residents who need them.

8.3.2 Unmet Need for Health Services

The measurement of the percentage of unmet need for health services is useful to assess the coverage of residents who should seek treatment when sick, but in reality, do not seek treatment because they cannot pay for the cost of treatment or transportation, or because there are no transportation facilities, or the long waiting time for services makes it difficult to seek treatment. To measure this, the number of residents who have health complaints but do not seek treatment at health facilities is divided by the number of sick residents, then multiplied by 100%. If the result is 0%, then this SDGs Desa target is achieved. The formula used is as follows:

$$\%\ \textit{Unmet need for health services} = \frac{\sum \text{Sick village residents who do not seek treatment at health facilities}}{\sum \text{Sick village residents}} \times 100\%$$

The target for the unmet need for health services is set at 0%. This target is measured using individual indicators, namely the number of sick residents who did not seek treatment at health facilities in the past year and the number of residents who were sick in the past year. The activities that the village government can undertake to achieve this target include establishing village clinics, providing basic medicines, and conducting training and development in the use of family medicinal plants.

8.3.3 Percentage of Births at Health Facilities

To measure the percentage of births at health facilities, the number of mothers who gave birth at health facilities (such as hospitals, maternity hospitals, polyclinics, health centres, auxiliary health centres, village health posts, and community health posts) in the past year is divided by the number of mothers who gave birth in the past year, then multiplied by 100%. If the result is 100%, then this SDGs Desa target is achieved. The formula used is as follows:

$$\%\ \textit{Childbirth in healthcare facilities} = \frac{\sum \text{Mothers who gave birth in healthcare facilities}}{\sum \text{Mothers who gave birth}} \times 100\%$$

The target percentage of births in health care facilities is set at 100%. This target is measured using individual indicators, namely the number of mothers giving birth and the number of mothers giving birth in health care facilities. To achieve this target, actions that can be taken are improving the condition of health facilities suitable for childbirth, encouraging and campaigning for childbirth to be carried out in adequate health facilities, providing cost assistance for health facilities to mothers giving birth who are from poor families, and improving or enhancing village health and childbirth clinic facilities.

8.3.4 Percentage of Births by Skilled Health Personnel

To measure the percentage of births by skilled health personnel, the number of mothers giving birth with the help of skilled health workers (such as midwives, trained health workers, and doctors) in the last year is divided by the number of mothers giving birth in the last year, then multiplied by 100%. If the result is 100%, then this SDGs Desa target is achieved. The formula used is as follows:

$$\%\ Childbirth\ assisted\ by\ skilled\ health\ personnel$$
$$= \frac{\sum \text{Mothers who gave birth assisted by skilled health personnel}}{\sum \text{Mothers who gave birth}} \times 100\%$$

The target percentage of births by skilled health personnel is set at 100%. This target is measured using individual indicators, namely the number of mothers giving birth and the number of mothers giving birth with the help of skilled health personnel. Activities to improve medical services with professional personnel for mothers giving birth are providing financial assistance for health facilities to mothers giving birth who are from poor families; and conducting health personnel training in villages that do not yet have skilled health personnel to assist in childbirth.

8.3.5 Maternal Mortality Rate Per 100,000 Live Births

To measure the maternal mortality rate, the number of maternal deaths in the last stage of pregnancy or childbirth in the last year is divided by the number of live births in the last year, then multiplied by 100,000. If the result is 0, then this SDGs Desa target is achieved. The formula used is as follows:

$$Maternal\ Mortality\ Rate$$
$$= \frac{\sum \text{Maternal Mortality during pregnancy or childbirth}}{\text{Number of live births}} \times 100.000$$

The target maternal mortality rate per 100,000 live births is set at 0. This target is measured using individual indicators, namely the number of maternal deaths during pregnancy or childbirth and the number of live births. The activities that can be carried out to achieve this target include providing healthy food to improve nutrition for pregnant, postpartum, and breastfeeding mothers; providing financial assistance to health facilities for mothers giving birth who are from low-income families; and also providing healthcare cost assistance post-childbirth for these mothers.

8.3.6 Infant Mortality Rate Per 1,000 Live Births

To measure the infant mortality rate, the number of infant deaths before reaching the age of one year during the last year is divided by the number of live births in the last year, then multiplied by 1,000. If the result is 0, then this SDGs Desa target is achieved. The formula used is as follows:

$$\textit{Infant Mortality Rate} = \frac{\sum \text{Infant mortality before the age of 1 year}}{\text{Number of live births}} \times 1000$$

The target infant mortality rate per 1000 live births is set at 0. This target can be measured using individual indicators, namely the number of infant deaths and births. To achieve this target, activities can be carried out such as: providing healthy food to improve nutrition for breastfeeding mothers from low-income families, providing infant milk support for newborns from low-income families, and offering free support activities periodically for babies and breastfeeding mothers to meet with health workers.

8.3.7 Percentage of Households with Complete Basic Immunisation for Babies

To measure complete basic immunisation in children aged 12–23 months for the 40% lowest-income families, the number of children aged 12–23 months old who receive complete basic immunisation with a criterion of 40% lowest-income families divided by the number of children aged 12–23 months, then multiplied by 100%. If the result reaches 100%, then the SDGs Desa target is achieved. The formula used is as follows:

$$\%\ \textit{Complete basic immunisation for children aged}\ 12 - 23\ \textit{months} = \frac{\sum \text{Children aged } 12 - 23 \text{ months receiving complete basic immunisation}}{\sum \text{Children Aged } 12 - 13 \text{ months}} \times 100\%$$

The target percentage of households that have complete basic immunisation for infants is set at 100%. This target is measured using individual indicators, namely the number of immunisations in children aged 12–23 months and the number of children aged 12–23 months who receive complete immunisation. To achieve this target, the village government can carry out activities such as conducting socialisation and immunisation campaigns, as well as organising educational activities and the provision of basic immunisation by health workers.

8.3.8 HIV Disease Prevalence

To measure the prevalence of HIV disease, the number of people who have suffered from HIV in the past year is divided by the total population, then multiplied by 100%. If the result is 0%, then this SDGs Desa target is achieved. The formula used is as follows:

$$Prevalence\ of\ HIV\ Diseases = \frac{\sum \text{HIV cases over the past year}}{\sum \text{Population}} \times 100\%$$

The target for HIV prevalence in the adult population is set at 0%. This target is measured using individual indicators, namely the number of family members and the number of family members with HIV. To suppress HIV prevalence in villagers, the village government can carry out activities, such as campaigns to promote clean and healthy living behaviours (Perilaku Hidup Bersih dan Sehat/PHBS), as well as efforts to prevent infectious diseases, sexually transmitted diseases, and HIV.

8.3.9 Prevalence of Tuberculosis

To measure the prevalence of tuberculosis, the number of people who suffered from tuberculosis in the last year is divided by the population, then multiplied by 100%. If the result is 0%, then this SDGs Desa target is achieved. The formula used is as follows:

$$Prevalence\ of\ Tuberculosis\ Disease = \frac{\sum \text{Tuberculosis cases over the past year}}{\sum \text{Population}} \times 100\%$$

The target prevalence of tuberculosis (TB) per 100,000 village population is set at 0%. This target is measured using individual indicators, namely the number of family members and the number of family members suffering from tuberculosis (TB). To achieve this target, the village government can carry out activities, such as organising campaigns and promoting clean and healthy living behaviours (Perilaku Hidup Bersih dan Sehat/PHBS), as well as making efforts to prevent and manage tuberculosis (TB).

8.3.10 Prevalence of High Blood Pressure

To measure the prevalence of high blood pressure, the number of people suffering from high blood pressure in the last year is divided by the general population, then multiplied by 100%. If the result is 0%, then this SDGs Desa target is achieved. The formula used is as follows:

$$Prevalence\ of\ High\ Blood\ Pressure\ Disease = \frac{\sum \text{High Blood Pressure cases over the past year}}{\sum \text{Population}} \times 100\%$$

The target for the prevalence of high blood pressure is set at 0%. This target is measured using individual indicators, namely the number of family members and the number of family members who have high blood pressure. Activities that can be carried out as prevention and emergency management related to high blood pressure cases are: conducting socialisation of high blood pressure prevention and actions taken in first aid.

8.3.11 Prevalence of Obesity in Population Aged Over 18 Years

To measure the prevalence of obesity, the number of obese people in the past year is divided by the total population, then multiplied by 100%. If the result is 0%, then this SDGs Desa target is achieved. The formula used is as follows:

$$Prevalence\ of\ Obesity = \frac{\sum \text{Obesity cases over the past year}}{\sum \text{Population}} \times 100\%$$

The target for obesity prevalence among the population aged over 18 years is set at 0%. This target is measured using individual indicators, namely the number of family members aged over 18, as well as their height and weight. Activities that the village government can carry out to manage and prevent obesity in village residents are: building simple physical exercise facilities in the village, providing support for healthy food and free nutrition consultation for village residents, and organising a healthy exercise day every week for village residents.

8.3.12 Drug Prevalence

To measure drug prevalence, the number of residents who have abused drugs in the past year is divided by the number of residents, then multiplied by 100%. If the result is 0%, then this SDGs Desa target is achieved. The formula used is as follows:

$$Drug\ Prevalence = \frac{\sum \text{Population with drug misuse over the past year}}{\sum \text{Population}} \times 100\%$$

The target for prevalence of narcotic abuse in the village is set at 0%. This target is measured using individual indicators, namely the number of family members, as well as the number of family members who are narcotic abusers. Activities that the village government can do, among others, are: campaigns about the dangers of narcotics, volunteer training, activists or anti-narcotics task forces, and opening reporting services for activities regarding Narcotic Abuse and Illicit Trafficking and Narcotic Precursors (Pencegahan, Pemberantasan Penyalahgunaan dan Peredaran Gelap Narkoba/P4GN).

8.3.13 Neighbourhood Units with Malaria Elimination

To measure the Neighbourhood Units/hamlets (Rukun Tetangga/RT) with malaria elimination, the number of hamlets (Rukun Tetangga/RT) with the condition of no malaria cases is divided by the number of hamlets (Rukun Tetangga/RT) that exist in the village, then multiplied by 100%. If the result is 100%, then this SDGs Desa target is achieved. The formula used is as follows:

$$Neighbourhood\ Units\ (RT)\ with\ Malaria\ Elimination = \frac{\sum \text{Neighbourhood Unit (RT) with no Malaria cases}}{\sum \text{Neighbourhood Unit (RT)}} \times 100\%$$

The target for hamlets (Rukun Tetangga/RT) with malaria elimination is set at 100%. This target is measured using the indicators of Neighbourhood Units/ hamlets (Rukun Tetangga/RT) and individuals, namely the number of hamlets (Rukun Tetangga/RT) with no malaria cases and the number of family members suffering from malaria. Activities that can be undertaken by the village government for the prevention and emergency management of malaria cases include: conducting fogging in the village, organising campaigns and socialisation related to the prevention of malaria and its management, as well as providing financial support for the treatment of malaria for the less fortunate.

8.3.14 Neighbourhood Units with Leprosy Elimination

To measure Neighbourhood Units/hamlets (Rukun Tetangga) with leprosy elimination, the number of hamlets (Rukun Tetangga/RT) found to have no cases of leprosy is divided by the number of hamlets (Rukun Tetangga/RT) in the village, then multiplied by 100%. If the result is 100%, then this SDGs Desa target is achieved. The formula used is as follows:

$$\textit{Neighbourhood Units (RT) with Leprosy Elimination} = \frac{\sum \text{Neighbourhood Unit (RT) with no Leprosy cases}}{\sum \text{Neighbourhood Unit (RT)}} \times 100\%$$

The target for RTs with leprosy elimination is set at 100%. This target is measured using the Neighbourhood Units/hamlets (Rukun Tetangga/RT) and individual indicators, namely the number of hamlets (Rukun Tetangga/RT) found to have no leprosy cases, and the number of family members suffering from leprosy. Activities that can be carried out to encourage the community to live healthily and avoid activities that can result in the transmission of leprosy, are, among others: conducting campaigns and socialisation to educate the community about the prevention and management of leprosy, as well as providing financial support for the management of leprosy for the less fortunate.

8.3.15 Neighbourhood Units with Filariasis Elimination

To measure the Neighbourhood Units/hamlets (Rukun Tetangga/RT) with Filariasis (Elephantiasis) elimination, the number of hamlets (Rukun Tetangga/RT) where no cases of Filariasis (Elephantiasis) are found is needed, divided by the total number of hamlets (Rukun Tetangga/RT) in the village, then multiplied by 100%. If the result is 100%, then this SDGs Desa target is achieved. The formula used is as follows:

$$\textit{Neighbourhood Units(RTs) with Filariasis(Elephantiasis) elimination} = \frac{\sum \text{Neighbourhood Unit (RT) with no Filariasis (Elephantiasis) cases}}{\sum \text{Neighbourhood Unit (RT)}} \times 100\%$$

The target number of hamlets (Rukun Tetangga/RT) with filariasis (elephantiasis) elimination is set at 100%. This target is measured using the indicators of hamlets (Rukun Tetangga/RT) and individuals, namely the number of hamlets (Rukun Tetangga/RT) with no filariasis (elephantiasis) cases, and the number of family members suffering from filariasis (elephantiasis). Activities that can be carried out for prevention and emergency management of elephantiasis cases include: organising campaigns and socialisation to educate village residents about the prevention and handling of filariasis disease (elephantiasis), and providing financial support for the treatment of filariasis (elephantiasis) for underprivileged village residents.

8.3.16 Smoking Percentage in Population Aged ≤ 18 Years

To measure the smoking percentage in the population aged ≤ 18 years the number of people aged ≤ v18 years who smoke is divided by the number of people aged ≤

v18 years, then multiplied by 100%. If the result is 0%, then this SDGs Desa target is achieved. The formula used is as follows:

$$\% \, Smoking \; in \; population \; aged \leq 18 \; years = \frac{\sum \text{Population aged} \leq 18 \text{ years who smoke}}{\sum \text{Population aged } \leq 18 \text{ years}} \times 100\%$$

The target percentage of smokers in the population aged under 18 years is set at 0%. This target is measured using family and individual indicators, namely the number of family members aged ≤ 18 years, and the number of family members under 18 years old who smoke. Activities can be carried out by the village government, one of which is to prohibit and discourage smoking for children under the age of 18 years through a campaign on the dangers of smoking for the community.

8.3.17 Neighbourhood Units that Have Integrated Health Posts Providing Mental Health Support

To measure the increase in Neighbourhood Units/hamlets (Rukun Tetangga/RT) that have Integrated Health Posts (Posyandu) providing mental health services, the number of hamlets (RT) that have Posyandu providing mental health services in a certain year is needed, then subtract the number of hamlets (RT) that had Posyandu providing mental health services in the previous year, and then divide by the number of hamlets (RT) that have Posyandu providing mental health services in the previous year, then multiply by 100%. If the result is positive (indicating an increase), then the SDGs Desa target has been achieved. The formula is as follows:

$$Increasing \; the \; number \; of \; NeighbourhoodUnit(RT) \; with \; Posyandu \; providing \; mental \; health \; services = \frac{\sum \begin{array}{l} NeighbourhoodUnit(RT) \; with \; Posyandu \; providing \; mental \; health \; services \; in \; a \; certain \; year - \\ NeighbourhoodUnit(RT) \; with \; Posyandu \; providing \; mentalh \; health \; services \; in \; the \; previous \; year \end{array}}{\text{Neighbourhood Unit (RT) with Posyandu providing mental health services in the previous year}} \times 100\%$$

The target for hamlets (RT) with Posyandu providing mental health services is set to increase. This target is measured by the number of Posyandu conducting mental health efforts in the last year, as well as the number of Posyandu conducting mental health efforts in the previous year in the village. Activities that the village government can do to increase the number of Posyandu supporting mental health are: proposing to the government more training of mental health workers, and, increasing the construction of Community Health Centres equipped for mental health services in villages/districts.

8.3.18 Victims of Drug Abuse Receiving Social Rehabilitation in Shelters According to Service Standards

To measure the percentage of victims of drug abuse (narcotics, psychotropics, and other addictive substances) who receive social rehabilitation in shelters according to service standards, the number of victims of drug abuse who receive social rehabilitation in shelters according to service standards is needed, divided by the number of victims of drug abuse, then multiplied by 100%. If the result is 100%, then this SDGs Desa target is achieved. The formula used is as follows:

$$\textit{Victims of Drug Abuse receiving Social Rehabilitation in Facilities according to Service Standards} = \frac{\sum \text{Victims of Drug Abuse receiving Social Rehabilitation in Facilities according to Service Standards}}{\sum \text{Victims of drug abuse}} \times 100\%$$

The target for victims of drug abuse who receive social rehabilitation in shelters according to service standards is set at 100%. This target is measured using individual indicators, namely the number of family members who are victims of drug abuse, and the number of victims of drug abuse who receive social rehabilitation in the shelter. Efforts that can be made by the village government, include identifying village residents who are victims of drug abuse, encouraging them to undergo rehabilitation according to standards, holding socialisation about the importance of rehabilitation at standard social shelters for village residents who are victims of drug abuse, and coordinating with the district government and related agencies in managing and supporting the rehabilitation of drug abusers in the village.

8.3.19 Victims of Death and Serious Injury Due to Traffic Accidents

To measure the percentage of victims of death and serious injury due to traffic accidents, the number of victims of death and serious injury due to traffic accidents is divided by the number of victims of death and serious injury, then multiplied by 100%. If the result is 0%, then this SDGs Desa target is achieved. The formula used is as follows:

$$\%\textit{Victims of Death and Serious Injury due to Traffic Accidents} = \frac{\sum \text{Victims of death and serious injury due to traffic accidents}}{\sum \text{Victims of death and serious injury}} \times 100\%$$

The target number of fatalities and serious injuries due to traffic accidents is set at 0%. This target is measured using individual indicators, namely the number of family members who died due to traffic accidents over the past year, and the number of family members who suffered serious injuries due to traffic accidents over the past year. Activities that the village government can undertake to achieve this target include: expediting first aid and facilitating greater access to health services for accident victims, constructing a medical centre for emergency first aid, and procuring vehicles or boats for use as village ambulances.

8.3.20 Prevalence of Short-Term Contraceptive Use Among Married People of Productive Age

Short-term contraceptive methods consist of injections, condoms, and pills. To measure the prevalence of short-term contraceptive use among married people of productive age (18–49 years), the number of married people of productive age (18–49 years) who use short-term contraception is divided by the number of married people of productive age (18–49 years), then multiplied by 100%. If the result reaches 100%, then the SDGs Desa target is achieved. The formula used is as follows:

$$\%\ Short-Term\ Contraceptive\ Use\ among\ Married\ People\ of\ Productive\ Age$$
$$= \frac{\sum \text{Married people of productive age using short} - \text{term contraceptive}}{\sum \text{Married people of productive age } (18-49 \text{ year})} 100\%$$

The target prevalence of short-term contraceptive use among married people of productive age (18–49 years) is set at 100%. This target is measured using individual indicators, namely the number of family members of productive age (18–49 years) who are married, and the number of family members using short-term contraceptives. Activities that can be carried out to introduce and socialise the use of short-term contraceptives to the community include: organising campaigns and socialisation about health services, population, reproductive health and family planning, as well as conducting training for population cadres, family planning, and family development.

8.3.21 Usage Rate of Modern Long-Term Contraceptive Methods in Married People of Productive Age

Modern contraceptive methods consist of female sterilisation, male sterilisation, pills, coil/IUD, contraceptive injection, contraceptive implant, condoms, lactational amenorrhea method (LAM). To measure the usage rate of Long-Term Contraceptive

Methods (Metode Kontrasepsi Jangka Panjang/MKJP) in a modern way in married people of productive age (ages 18–49 years) requires the number of married people of productive age (ages 18–49 years) who use Long-Term Contraceptive Methods in a modern way divided by the number of married people of productive age (ages 18–49 years), then multiplied by 100%. If the result reaches 100%, then the SDGs Desa target is achieved. The formula used is as follows:

$$\begin{aligned}&\%\ \textit{Usage of Modern Long}-\textit{Term Contraceptive Methods}\\&\textit{in Married People of Productive Age}\ (18-49\ \textit{years old})\\&=\frac{\sum \begin{array}{c}\textit{Married people of productive age}\\ \textit{using modern long}-\textit{term contraceptive methods}\end{array}}{\sum \text{Married people of productive age}}\times 100\%\end{aligned}$$

The target rate of use of the long-term contraceptive method, using modern methods for married people of reproductive age (18–49 years) is set at 100%. This target is measured using individual indicators, namely the number of family members of reproductive age (18–49 years) who are already married, and the number of family members who use long-term contraception. Activities that can be carried out as an effort to introduce and socialise the use of long-term contraceptive devices to villagers, are: organising campaigns and socialisation about health services, population, reproductive health and family planning; optimising family planning through mentoring and guidance to prospective family planning participants; as well as improving access and quality of family planning services.

8.3.22 Birth Rate in Adolescents Aged 15–19 Years

To measure the Age Specific Fertility Rate (ASFR) for the number of births in the group of adolescent females aged 15–19 years can be done by adding up the births to adolescents aged 15–19 years divided by the number of adolescents aged 15–19 years, then multiplied by 100%. If the result is 0%, then the SDGs Desa target is achieved. This measurement uses the following formula:

$$\begin{aligned}&\textit{Age Specific Fertility Rate}\ (\textit{ASFR})\\&=\frac{\sum \text{Birth Rate in Adolescents}\ (15-19\ \text{years old})}{\sum \text{Adolescents aged}\ 15-19\ \text{years}}\times 100\%\end{aligned}$$

The target for the adolescent birth rate for ages 15–19 years (ASFR) is set to 0%. This target is measured using individual indicators, namely the number of family members aged 15–19 years, and the number of family members aged 15–19 years who gave birth in the last year. Activities carried out to reduce the number of underage marriages and pregnancies include: conducting reproductive health and family planning campaigns, running campaigns to prevent underage marriage, and formulating regulations on the age of child marriage in specific laws.

8.3.23 *Total Fertility Rate (TFR)*

Measuring the TFR or total fertility rate can be done by taking the total births in women of childbearing age (15–49 years) divided by the number of women of childbearing age (15–49 years). If the result is below 1.5, then the SDGs Desa target is achieved. This measurement uses the following formula:

$$Total\ Fertility\ Rate\ (TFR) = \frac{\sum \text{Births in women of childbearing age } (15-49 \text{ years})}{\sum \text{Women of childbearing age } (15-49 \text{ years})}$$

The target for the Total Fertility Rate (TFR) is set to be below 1.5. This target is measured using individual indicators, such as the age of family members, the number of female family members, and the number of births in the family in the last year. Activities that can be carried out to increase health support for mothers giving birth include: providing healthy food to improve nutrition for pregnant, postpartum, and breastfeeding mothers, providing financial assistance to health facilities for mothers giving birth who are from low-income families, and providing postnatal health care cost assistance for mothers giving birth who are from low-income families.

8.4 Operational Definitions for a Healthy and Prosperous Village

Key words related to a healthy and prosperous village, SDGs Desa 3 Are Found in Table 8.1.

Table 8.1 Operational Definitions for a Healthy and Prosperous Village

	Keyword	Operational definition
1	Age Specific Fertility Rate/ ASFR	The number of births per 1,000 women in a certain age group
2	Maternal Mortality Rate (MMR)	The number of female deaths during pregnancy or within 42 days of termination of pregnancy, regardless of the duration and place of delivery, caused by the pregnancy or its management, and not by other causes, per 100,000 live births
3	Baduta/Bawah Dua Tahun (Under Two Years)	One of the human age periods after infancy with an age range from one to two years, or usually with the age calculated in months: 12–24 months
4	Balita/Bawah Lima Tahun (Under Five Years)	One of the human age periods after infancy with an age range from three to five years, or usually with the age calculated in months: 36–60 months
5	Batita/Bawah Tiga Tahun (Under Three Years)	One of the human age periods after infancy with an age range from two to three years, or usually with the age calculated in months: 24–36 months
6	Filariasis (elephantiasis)	Limb swelling due to infection by the filaria worm
7	Nutrition	Organic substances required by organisms for normal body functions, growth, and health maintenance
8	HIV (human immunodeficiency virus)	A virus that damages the immune system by infecting and destroying CD4 cells. Untreated HIV infection will develop into a serious condition called AIDS (Acquired Immune Deficiency Syndrome) where the body's ability to fight infection is completely lost
9	Child immunisation	The administration of a vaccine into a child's body to provide immunity against a particular disease
10	Leprosy	A chronic bacterial infection that attacks skin tissue, peripheral nerves, and the respiratory tract
11	Malaria	A contagious infectious disease spread through mosquito bites
12	NAPZA (Narcotics, Psychotropics, and Addictive Substances)	Narcotics are substances or drugs derived from plants or non-plants, whether synthetic or semi-synthetic, that can cause a decrease and change in consciousness, reduce or eliminate pain, and can cause addiction Psychotropics are substances or drugs, whether natural or synthetic non-narcotics, that have psychoactive effects through selective influence on the central nervous system causing behavioural changes and characteristic changes in mental activity Addictive substances are materials or substances that have psychoactive effects outside of narcotics and psychotropics
13	Obesity	A chronic condition due to a very high accumulation of fat in the body
14	Childbirth	A state/process in which a child is expelled from the mother's body

(continued)

Table 8.1 (continued)

	Keyword	Operational definition
15	Polyclinic	A general medical centre (not for care or overnight patients)
16	Polindes (Village Maternity Hut)	A form of participation or community involvement in providing a place for childbirth assistance and mother and child health services, including the Family Planning (KB) which is located in the village
17	Village Health Post (Poskesdes)	A community-based health effort (UKBM) established in the village to bring/provide basic health services to the village community
18	Community Health Centre (Puskesmas)	A health service facility under the supervision of the District/City Health Office that carries out public health efforts and individual health efforts at the first level, prioritising promotive and preventive efforts, to achieve the highest degree of public health
19	Auxiliary Community Health Centre (Pustu)	A network of Community Health Centre services that provide permanent health services at a location within the Community Health Centre's working area. Pustu is an integral part of the Community Health Centre, which must be regularly supported and managed by the Community Health Centre
20	Disease History and Abnormalities	A record of having experienced a certain abnormal condition that negatively affects the structure or function of part or all of a living creature's body and is not caused by any external injury
21	Hospital	A health service institution that provides comprehensive individual health services that provide inpatient, outpatient, and emergency services
22	Maternity Hospital (RSB)	A hospital that specifically serves the examination of pregnant women, women about to give birth and the health of children under the age of five
23	Medicinal Plants	Plants that have been identified and known based on human observation to have compounds beneficial for preventing and curing diseases, performing certain biological functions, to preventing insect and fungus attacks
24	Tuberculosis (TB)	A directly contagious disease caused by the TB germ (Mycobacterium tuberculosis)
25	Total Fertility Rate (TFR)	The average number of children a woman will give birth to during her reproductive period
26	Unmet Need for Health Services	Population who do not seek treatment although they have health complaints that disrupt their daily life

References

Ministry of Health. 2017. *Healthy Indonesia Program with Family Approach.* Available at: https://www.kemkes.go.id/article/view/17070700004/program-indonesia-sehat-dengan-pendekatan-keluarga.html

Ministry of Health. 2018. *Main Results of RISKESDAS.* Available at: https://kesmas.kemkes.go.id/assets/upload/dir_519d41d8cd98f00/files/Hasil-riskesdas-2018_1274.pdf

Chapter 9
SDGs Desa 4: Quality Village Education

9.1 Concept of Quality Village Education

There are several development goals consisting of income for poor villagers, ensuring access to basic services for villagers, and protecting all villagers from all forms of disaster. In order to achieve the goal of increasing income for poor villagers, one method is to improve the quality of the village's human resources (HR). Education is a key variable in driving upthe quality of human resources. Education can be said to be a social investment for each villager to achieve success in the future. Through education, an individual will increase their competitiveness. When social investment through the education of each of these individuals is aggregated, collected, and connected to the national level, then it will accumulate into an improvement in the quality of human resources of Indonesia, increasing the competitiveness of the Indonesian nation.

Quality education is education that focuses on developimg the whole child—social, emotional, mental, physical, and cognitive development—regardless of gender, race, ethnicity, social economic status, or geographical location, by preparing children for life, not just for exams (Huffpost, n.d.). Quality education also gives all learners the skills they need to be economically productive, develop sustainable livelihoods, contribute to a peaceful and democratic society and improve individual well-being (VVOB, n.d.).

Education not only plays a role in creating young generations as *agents of change* that bring about change. Education must also be able to shape young generations into *agents of producers* who are capable of creating real change. Education must be able to shape the character and mindset of learners, education must also be able to encourage the creativity and innovative power of the nation's children. Education must give birth to a generation that can make a significant contribution and play an active role in sustainable national development.

In 2020, Indonesia ranked 54th out of 73 countries with the best education in the world (U.S. News and World Report 2020). Meanwhile, in Southeast Asia, Indonesia

A. H. Iskandar, *The Methods of Measuring the National Sustainable Development Goals in Indonesia*, Engaging Indonesia, https://doi.org/10.1007/978-981-96-0927-7_9

ranks fourth after Singapore, Malaysia, Thailand, and the Philippines. As can be seen in Fig. 9.1, the reading score of Indonesians follows a *hump- shaped* curve, meaning it is increasingly negative from year to year (PISA 2018).

Nationally, the literacy rate has increased relatively from 2015 to 2020, with a figure of 96% (BPS, 2020). However, there is a gap between the literacy rate of rural areas and the urban literacy rate (Fig. 9.2). In 2020, the literacy rate of rural areas was 93.64%, while the urban literacy rate reached 97.81%. Therefore, achieving this SDGs Desa 4 is very important in realizing Sustainable Development Goals.

Furthermore, the average length of schooling in Indonesia is still very low. In 2019, the average length of schooling for the Indonesian population was only 8.34 years, which then saw a slight increase in 2020 to 8.48 years. This means that, on average,

Fig. 9.1 Indonesia's reading score in PISA, 2000–2018

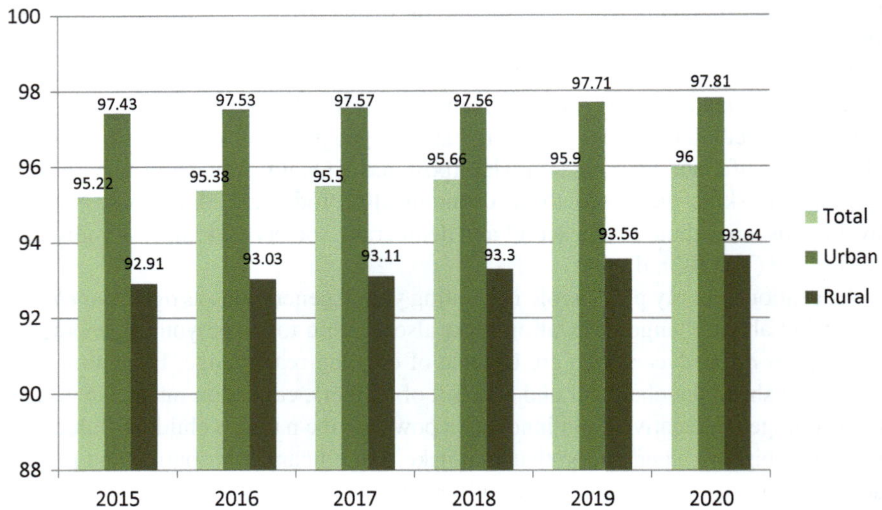

Fig. 9.2 Literacy rate of population aged 15 years and over by place of residence in Indonesia, 2015–2020

Indonesians are unable to complete their first stage of secondary education, and may drop out in the final year of their first stage of secondary education (BPS 2020a).

Through education, the quality and competitiveness of villages in Indonesia can continue to progress, and the Indonesian nation can compete with other countries around the world. Therefore, village governments, together with supra-village entities, must ensure the availability and easy accessibility of quality education services for village residents.

Therefore, to achieve the goal of SDGs Desae 4, Quality Village Education, what the village government needs to do, with support from supra-village entities, is to provide village residents with access to accredited education services and access to Islamic boarding school education institutions, and to ensure the availability of quality and affordable education services for village residents.

In addition, this goal focuses on the availability of skills education services for village residents, pre-school education services, non-formal education, and the availability of reading parks or village libraries.

9.2 Quality Village Education Targets

To achieve the SDGs Desa goal 4: Quality Village Education, the following 16 targets to be achieved by 2030 have been set:

1. The percentage of access to accredited primary school/equivalent (SD/MI/sederajat) reaching at least B is 100%
2. The percentage of access to accredited junior high school/equivalent (SMP/MTs/sederajat) reaching at least B is 100%
3. The percentage of access to accredited senior high school/senior vocational school/equivalent (SMA/SMK/MA/sederajat) reaching at least B is 100%
4. The percentage of access to Islamic boarding schools is 100%
5. The Gross Enrolment Rate (GER) for primary school/equivalent (SD/MI/sederajat) reaches 100%
6. The Gross Enrolment Rate (GER) for junior high school/equivalent (SMP/MTs/sederajat) reaches 100%
7. The Gross Enrolment Rate (GER) for senior high school/equivalent (SMA/SMK/MA/sederajat) reaches 100%
8. The average length of schooling for the population aged over 20 years reaches 12 years
9. The Gross Enrolment Rate (GER) for children attending Early Childhood Education (Pendidikan Anak Usia Dini/PAUD) and kindergarten (Taman Kanak-Kanak/TK) reaches 100%
10. The Gross Enrolment Rate (GER) for Higher Education reaches 100%
11. The ratio of the Net Enrolment Rate (NER) for females/males in primary school/Package A/equivalent (SD/MI/A that is gender equivalent reaches 100%
12. The ratio of the Net Enrolment Rate (NER) for females/males in junior high school/Package B/equivalent (SMP/MTs/Paket B/sederajat) that is gender equivalent reaches 100%

13. The ratio of the Net Enrolment Rate (NER) for females/males in senior high school/Package C/equivalent (SMA/SMK/MA/Paket C/sederajat) that is gender equivalent reaches 100%
14. The literacy rate for Latin and non-Latin scripts for the population aged over 15 years reaches 100%
15. The literacy rate for Latin and non-Latin scripts for the adult population aged 15-59 years reaches 100%
16. Availability of Community Reading Parks or libraries.

9.3 Measurement and Analysis of Quality Village Education

9.3.1 Percentage of Access to Primary Schools or Equivalent with at Least B Accreditation

To measure the percentage of children's access to accredited primary education/equivalent (SD/MI/sederajat) with at least a B accreditation, data on the nearest accessible primary education for each household is needed. Each household gives a value of 1 if access to a B-accredited primary school/equivalent (SD/MI/sederajat) is easy and gives a value of 0 if access to a B-accredited primary school/equivalent (SD/MI/sederajat) is difficult. Then the number of these accessibilities is divided by the number of households, then multiplied by 100%. If the result is 100%, then the SDGs Desa target is achieved. The formula used is as follows:

$$\%\text{Nearby primary schools with at least B accreditation}$$
$$= \frac{\sum \text{Accessibility of nearby primary schools or equivalent with B accreditation}}{\sum \text{Neighbourhood Units}} \times 100\%$$

The target percentage of access to accredited primary schools/equivalent (SD/MI/sederajat) with at least a B accreditation is set at 100%. The village indicators used are the number of primary schools/equivalent (SD/MI/sederajat), the number of A and B accredited primary schools/equivalnt (SD/MI/sederajat), the number of teachers, and the number of students in the village. The individual indicators used are the number of children attending primary schools/equivalent (SD/MI/sederajat) and their school accreditation. Activities that can be carried out to increase the villagers' access to primary schools/equivalent (SD/MI/sederajat) include: building student dormitories/temporary student homes; creating free Wi-Fi facilities for student learning in certain spots within the village, providing school buses/student pick-up vehicles, and introducing additional educational facilities in the village, such as libraries and information and technology hubs.

9.3.2 Percentage of Access to Junior High Schools with at Least B Accreditation

To measure the percentage of children's access to junior high schools/equivalent (SMP/MTs/sederajat) with at least a B accreditation, data on the nearest accessible junior high school education for each household is needed. Each household gives a value of 1 if access to junior high schools/equivalent (SMP/MTs/serajat) with at least a B accreditation is easy and gives a value of 0 if access to junior high schools/equivalent (SMP/MTs/sederajat) with at least a B accreditation is difficult. Then, the number of these accessibilities is divided by the number of households, then multiplied by 100%. If the result is 100%, then the SDGs Desa target is achieved. The formula used is as follows:

$$\%\text{Nearby junior high schools with at least B accreditation}$$
$$= \frac{\sum \text{Accessibility of nearby junior high schools or equivalent with B accreditation}}{\sum \text{Neighbourhood Units (RT)}} \times 100\%$$

The target percentage for access to accredited junior high school/equivalent (SMP/MTs/sederajat) with at least a B accreditation is set at 100%. The village indicators used are the number of junior high school/equivalent (SMP/MTs/sederajat), the number of junior high school/equivalent (SMP/MTs/sederajat) with A and B accreditation, the number of teachers, and the number of students in the village. The individual indicators used are the number of children attending junior high school/equivalent (SMP/MTs/sederajat) and their school's accreditation. Activities that can be done to increase the accessibility of villagers to junior high school/equivalent (SMP/MTs/sederajat) include building student dormitories/temporary student homes, creating free Wi-Fi facilities for student learning in certain spots within the village, providing school buses/student pick-up vehicles, and adding educational facilities in the village, such as libraries and technology and information hubs. Other efforts include conducting school accreditation data collection and encouraging improvements to the Education Office of the Regency/City, as well as coordinating with the Regency Education Office regarding proposals for junior high school/equivalent (SMP/MTs/sederajat) educational needs in the village.

9.3.3 Percentage of Access to Senior High Schools or Equivalent with at Least B Accreditation

To measure the percentage of children's access to senior high schools/senior vocational high school/equivalent (SMA/MA/SMK/sederajat) with at least a B accreditation, data on the nearest accessible senior high school education for each household is needed. Each household gives a value of 1 if access to senior high school/senior vocational school/equivalent (SMA/MA/SMK/sederajat) with at least a B accreditation is easy and gives a value of 0 if access to senior high school/senior vocational school/equivalent (SMA/MA/SMK/sederajat) with at least a B accreditation is difficult. Then the number of accessibilities is divided by the number of households, then multiplied by 100%. If the result is 100%, then the SDGs Desa target is achieved. The formula used is as follows:

$$\%\text{Nearby senior high schools with at least B accreditation} = \frac{\sum \text{Accessibility of nearby senior high schools or equivalent with B accreditation}}{\sum \text{Neighbourhood Units/RT}} \times 100\%$$

The target percentage for access to accredited senior high school/senior vocational high school/equivalent (SMA/MA/SMK/equivalent) of at least B is set at 100%. The village indicators used are the number of high schools/vocational schools/equivalent, the number of A and B accredited senior high school/senior vocational school/equivalent (SMA/MA/SMK/sederajat), the number of teachers, and the number of students in the village. The individual indicators used are the number of children attending senior high school/senior vocational school/equivalent (SMA/MA/SMK/sederajat) and their school accreditation. Activities that can be done to increase the accessibility of villagers to senior high school/senior vocational school/equivalent (SMA/MA/SMK/sederajat) include building student dormitories/temporary student homes, creating free Wi-Fi facilities for student learning at certain spots within the village, providing school buses/student pick-up vehicles, and adding educational facilities in the village, such as libraries and information and technology hubs. Another effort is by conducting school accreditation data collection and encouraging to improve accreditation to the District/City Education Office and coordinating with the District Education Office regarding proposals for high school/vocational school/equivalent education needs in the village.

*9.3.4 Percentage Access to Islamic Boarding Schools (***Pesantren***)*

To measure the percentage of children's access to Islamic boarding school (*pesantren*) education, data on the number of nearest accessible Islamic boarding school for each household is needed. Each household gives a value of 1 if access to the Islamic boarding school is easy and gives a value of 0 if access to the Islamic boarding school is difficult. Then the number of these accessibilities is divided by the number of households, then multiplied by 100%. If the result is 100%, then the SDGs Desa target is achieved. The formula used is as follows:

$$\%\text{Access to Islamic Boardings Schools} = \frac{\sum \text{Access to the nearest Islamic Boarding Schools}}{\sum \text{Neighbourhood Units (RT)}} \times 100\%$$

The target percentage for Islamic boarding school or *pesantren* access is set at 100%. The village indicators used are the number of *pesantren* and the number of students in the village. The individual indicators used are the number of family members who are currently studying at the *pesantren*. Activities that can be carried out include building student dormitories/temporary student homes; creatinging free Wi-Fi facilities for student learning in certain spots within the village, providing school buses/student pick-up vehicles, and adding educational facilities in the village, such as libraries and information and technology hubs.

9.3.5 Gross Enrolment Rate (GER) for Primary School

The Gross Enrolment Rate (GER) is the ratio of the number of students, regardless of age, who are currently studying at a certain level of education to the number of people in the age group associated with that level of education. The GER is used to determine the level of participation of the population in general at a certain level of education. The GER is the simplest indicator to measure the absorption capacity of school-age population at each level of education, without considering age.

To measure the Gross Enrolment Rate (GER) for primary school/equivalent, data is needed on the number of children of various ages who are currently attending primary school/equivalent divided by the number of children who are in the 7–12 years age group in a particular year multiplied by 100%. If the result is 100%, then the SDGs Desa target is achieved. The formula used is as follows:

$$\%\ \text{GER for Primary School} = \frac{\sum \text{Primary or equivalent student}}{\sum \text{Children aged } 7-12 \text{ years}} \times 100\%$$

The target for the Gross Enrolment Rate (GER) for primary school/equivalent is set at 100%. The village indicator used is the number of primary school/equivalent students. The individual indicator used is the age (date of birth) of family members and the number of members who are currently in primary school/equivalent. Activities that can be carried out include: building Community Learning Activity Centres/CLAC (Pusat Kegiatan Belajar Masyarakat/PKBM) that provide Package A equivalency, forming village education cadres, conducting compulsory basic education campaigns, conducting out-of-school education such as skills training, and providing education support funds such as scholarships to students from poor populations.

9.3.6 Gross Enrolment Rate (GER) for Junior High School

To measure the Gross Enrolment Rate (GER) for junior high school/equivalent, data is needed on the number of children of various ages who are currently in Junior High School/equivalent in a given year. This number is then divided by the number of teenagers in the age group of 13–15 years in the same year and multiplied by 100%. If the result is 100%, then the SDGs Desa target is achieved. The following formula is used to measure the GER for Junior High School/equivalent.

$$\% \text{ GER for Junior High School} = \frac{\sum \text{Junior High School or equivalent student}}{\sum \text{Teenagers Aged } 13 - 15 \text{ years}} \times 100\%$$

The target for the Gross Enrolment Rate (GER) for junior high school/equivalent is set at 100%. The village indicator used is the number of junior high school/equivalent students. The individual indicator used is the age (date of birth) of family members and the number of members who are currently attending junior high school/equivalent. Activities that can be carried out include building a Community Learning Activity Centre that provides Package B equivalency, forming village community education cadres, conducting compulsory basic education campaigns, conducting out-of-school education such as skills training, and providing education support funds such as scholarships to students from poor populations.

9.3.7 Gross Enrolment Rate (GER) for Senior High School

To measure the Gross Enrolment Rate (GER) for senior high school/senior vocational high school/equivalent (SMA/MA/SMK/sederajat), data is needed on the number of children of various ages who are currently attending senior high school/senior vocational school/equivalent (SMA/MA/SMK/sederajat)—in a particular year. This number is then divided by the number of teenagers in the age group 16–18 years in a particular year, then multiplied by 100%. If the result is 100%, then the SDGs Desa target is achieved. The formula is as follows:

$$\% \text{ GER for Senior High School or Equivalent} = \frac{\sum \text{Senior High School or equivalent student}}{\sum \text{Teenagers Aged } 16 - 18 \text{ years}} \times 100\%$$

The target Gross Enrolment Rate (GER) for senior high school/senior vocational high school/equivalent (SMA/MA/SMK/sederajat) is set at 100%. The village indicator used is the number of senior high school/vocational school/equivalent students. The family/individual indicator used is the age (date of birth) of family members and the number of members who are currently attending senior high school/senior vocational high school/equivalent. Activities that can be carried out include building Community Learning Activity Centres that provide Package C equivalency, forming village education cadres, conducting compulsory basic education campaigns, conducting out-of-school education such as skills training, and providing education support funds such as scholarships to students from poor populations.

9.3.8 Average Length of Schooling for Population Aged Over 20 Years

The average length of schooling is the duration of the population's formal education in years. To calculate it, total data is needed regarding the length of schooling of the population aged over 20 years. This number is divided by the number of people aged 20 years and over. If the result reaches 12 years, then the SDGs Desa target is achieved. The formula is as follows:

$$\text{Average Length of Schooling} = \frac{\sum \text{Schooling length for the population aged} > 20 \text{ years}}{\sum \text{Population aged} > 20 \text{ years}}$$

The target for the average length of schooling for the population aged over 20 years is set at 12 years. The individual indicators used are the age (date of birth) of family members and the highest education level of family members. Activities that can be carried out include conducting a census of school dropouts, accommodating and facilitating school dropouts to participate in Community Learning Activity Centres (Pusat Kegiatan Belajar Masyarakat/PKBM, and providing financial assistance for the education of school dropouts.

9.3.9 Gross Enrolment Rate (GER) for Children Participating in Early Childhood Education and Kindergarten

To measure the Gross Enrolment Rate (GER) for early childhood education (Pendidikan Anak Usia Dini/PAUD) and kindergarten (Taman Kanak-Kanak) requires data on the number of children from various ages who, in a certain year, attend early childhood education (PAUD) and kindergarten. This number is divided by the number of children who, in a certain year, are in the age group 5–6 years, then multiplied by 100%. If the result is 100%, then the SDGs Desa target is achieved. The formula is as follows:

$$\%\ \text{GER for Early Childhood Education and Kindergarten} = \frac{\sum \text{Children participating in early childhood education and kindergarten}}{\sum \text{Children aged } 5 - 6 \text{ years}} \times 100\%$$

The target Gross Enrolment Rate (GER) for children attending early childhood education (PAUD) and kindergarten (Taman Kanak-Kanak) is set at 100%. The individual indicators used are the age (date of birth) of family members and the number of family members currently attending early childhood education (PAUD) and kindergarten (Taman Kanak-Kanak). Activities that can be carried out include building/

rehabilitating early childhood education (PAUD) and kindergarten (Taman Kanak-Kanak) facilities, providing learning facilities, creatinging facilities and infrastructure for children's play and creativity, forming early childhood education (PAUD) and kindergarten (Taman Kanak-Kanak) cadres in villages, improving the competence of early childhood education (PAUD) and kindergarten (Taman Kanak-Kanak) teachers, and providing incentives for early childhood education (PAUD) and kindergarten (Taman Kanak-Kanak) cadres/teachers.

9.3.10 Gross Enrolment Rate (GPR) for Higher Education

To measure the Gross Enrolment Rate (GER) for higher education, data is needed on the number of students of various ages in a particular year. This number is divided by the population in the same year in the age group 19–24 years, then multiplied by 100%. If the result is 100%, then the SDGs Desa target is achieved. The formula is as follows:

$$\%\ \text{GER for Higher Education} = \frac{\sum \text{College student}}{\sum \text{Population aged } 19 - 24 \text{ years}} \times 100\%$$

The target for the GER of higher education is set at 100%. The family/individual indicators used are the age (date of birth) of family members and the number of family members who are currently attending university or other higher education institution. Activities that can be carried out include providing village scholarships to village residents to attend higher education.

9.3.11 Ratio of Net Enrolment Rate of Female/Male in Primary School that is Gender Equal

The Net Enrolment Rate (NER) is the percentage of students with ages related to their level of education from the total population of the same age. The NER is used to identify school-age children who attend school at the standard age for the level of education. Like GER, NER is also an indicator of the absorption capacity of school-age population at each level of education. However, if compared to GER, NER is a better absorption indicator because it considers the participation of the standard age group population at the level of education that corresponds to that standard.

To measure the ratio of NER in primary school/equivalent data is needed on the number of children aged 7–12 years who are currently in school primary school/equivalent divided by the number of children aged 7–12 years then multiplied by 100%. If the result is 100%, then the SDGs Desa target is achieved. The formula used is as follows:

$$\% \text{ NER of Primary School} = \frac{\sum \text{Primary school student aged } 7-12 \text{ years}}{\sum \text{Children aged } 7-12 \text{ years}} \times 100\%$$

The target ratio of the NER of females/males in primary schools/equivalent that is gender equal is set at 100%. The individual indicators used are the age (date of birth) of family members, the gender of family members, the number of primary school children/equivalent based on gender. Activities that can be carried out include facilitating people with disabilities to get Special Education (SE), facilitating children in vulnerable conditions (neglected children) to get education, forming cadres for protection and empowerment of disabilities, conducting socialisation and advocacy of facilities and infrastructure that are friendly towards people with disabilities, managing rehabilitation activities for people with disabilities, empowering marginalised communities and people with disabilities, and developing Inclusive Villages.

9.3.12 Ratio of Net Enrolment Rate (NER) of Female/Male in Junior High Schools that is Gender Equal

To measure the ratio of the Net Enrolment Rate (NER) of junior high schools/equivalent data is needed on the number of teenagers aged 13–15 years who are currently in school junior high schools/equivalent. This number is divided by the number of teenagers aged 13–15 years, then multiplied by 100%. If the result is 100%, then the SDGs Desa target is achieved. The formula is as follows:

$$\% \text{ NER of Junior High School or Equivalent} = \frac{\sum \text{Junior High School Student aged } 13-15 \text{ years}}{\sum \text{Teenagers aged } 13-15 \text{ years}} \times 100\%$$

The target ratio of the Net Enrolment Rate (NER) of females/males in junior high school/Package B/equivalent (SMP/MTs/Paket B/sederajat) that is gender equivalent is set at 100%. The individual indicators used are the age (date of birth) of family members, gender of family members, number of junior high school/Package B/equivalent (SMP/MTs/Paket B/sederajat) school children based on gender. Activities that can be carried out include facilitating people with disabilities to obtain Special Education (Pendidikan Luar Biasa/PLB), facilitating children in vulnerable conditions (neglected children) to get education, forming cadres of protection and empowerment for the disabled, conducting socialisation and advocacy of facilities and infrastructure that are friendly to people with disabilities, managing rehabilitation activities for people with disabilities, empowering marginalised communities and people with disabilities, and developing Inclusive Villages.

9.3.13 Ratio of Net Enrolment Rate (NER) of Female/Male in Senior High School that is Gender Equal

To measure the ratio of the Net Enrolment Rate (NER) of senior high school/ senior vocational high school/equivalent (SMA/MA/SMK/sederajat), data on the number of teenagers aged 16–18 years who are currently attending senior high school/senior vocational school/equivalent (SMA/MA/SMK/sederajat) divided by the number of teenagers aged 16–18 years then multiplied by 100%. If the result is 100%, then the SDGs Desa target is achieved. The formula is as follows:

$$\%\ \text{NER of Senior High School} = \frac{\sum \text{Senior High School Student aged } 16-18 \text{ years}}{\sum \text{Teenagers aged } 16-18 \text{ years}} \times 100\%$$

The target ratio of the Net Enrolment Rate (NER) of females/males in senior high school/senior vocational high school/Package C that is gender equivalent is set at 100%. The individual indicators used are the age (date of birth) of family members, the gender of family members, the number of senior high school/senior vocational high school/Package C students based on gender. Activities that can be carried out include facilitating people with disabilities to get Special Education (Pendidikan Luar Biasa/PLB), facilitating children in vulnerable conditions (neglected children) to get education, forming protection and empowerment cadres for people with disabilities, conducting socialisation and advocacy of facilities and infrastructure that are friendly to people with disabilities, managing rehabilitation activities for people with disabilities, empowering marginalised communities and people with disabilities, and developing Inclusive Villages.

9.3.14 Literacy Rate in Latin and Non-Latin Scripts for Population Aged Over 15 Years

The literacy rate in Latin and non-Latin scripts depicts the population who have the ability to read and write simple sentences in Latin script and other scriptss (for example, Arabic and Javanese scripts). To measure the literacy rate in Latin and non-Latin scripts for the population aged over 15 years, data is needed on the number of people aged over 15 years who are literate in Latin and non-Latin scripts, This number is divided by the number of people aged over 15 years. If the result is 100%, then the SDGs Desa target is achieved. The formula is as follows:

$$\text{Literacy Rate in Latin and Non-Latin Scripts} = \frac{\sum \text{Population aged} > 15 \text{ years who are literate in latin and non} - \text{latin scripts}}{\sum \text{Population aged} > 15 \text{ years}} \times 100\%$$

The target for literacy in Latin and non-Latin among the population aged over 15 years is set at 100%. The indicators used are the individual's age (date of birth), the number of family members who are literate, and the number of family members who are illiterate. Activities that can be carried out include building community libraries/village libraries, developing literacy teaching programmes for the community who are not yet literate, procuring books/reading materials, building religious learning parks, and providing incentives for teachers/volunteers/facilitators in community learning centres.

9.3.15 Literacy Rate in Latin and Non-Latin Scripts Among the Population Aged 15–59 Years

To measure the literacy rate in Latin and non-Latin scripts among the population aged 15–59 years, data is needed on the number of people aged 15–59 years who are literate in Latin and non-Latin script. This number is divided by the total population aged 15–59 years. If the result is 100%, then the SDGs Desa target is achieved. The formula is as follows:

$$\text{Literacy Rate in Latin and Non - Latin Scripts} = \frac{\sum \text{Population aged } 15-59 \text{ years who are literate in latin and non}-\text{latin scripts}}{\sum \text{Population aged } 15-59 \text{ years}} \times 100\%$$

The target literacy rate in Latin and non-Latin scripts for the adult population aged 15–59 years is set at 100%. The individual indicators used are the age (date of birth) of family members, the number of family members aged 15–59 years who are literate, and the number of family members aged 15–59 years who are illiterate.

9.3.16 Availability of Community Reading Parks or Libraries

To measure the number of communities reading parks or libraries, data on the number of communities reading parks or libraries in the village is needed. If there is a community reading park or library in the village, then the SDGs Desa target is achieved. The formula is as follows:

$$\text{Availability of Community Reading Parks or Libraries} = \sum \text{Community Reading Parks or Libraries}$$

This target uses the Neighbourhood Units/hamlet (Rukun Tetangga/RT) indicator, namely the availability of a community reading park or library. Efforts to improve the quality of community reading parks in villages include creating a comfortable room,

Table 9.1 Operational definitions for quality village education

	Keyword	Operational definition
1	Gross Enrolment Rate (GER)	The Gross Enrolment Rate (GER) is the ratio of the total number of students, regardless of age, who are in school at a certain level of education, to the total population of the age group related to that level of education. GER indicates the level of participation of the general population at a certain level of education. GER is the simplest indicator to measure the absorption capacity of school-age population at each level of education, without considering age
2	Net Enrolment Rate (NER)	The Net Enrolment Rate (NER) is the percentage of students of an age that corresponds to their level of education from the total population of the same age. NER is used to discover school-age children who attend school at the standard level that corresponds to their age.. Compared to GER, NER is a better absorption indicator because NER looks at the participation of the standard age group at the level of education that corresponds to that standard
3	Latin literacy	The ability to identify, understand, translate, create, communicate and process the content of a series of texts found in printed materials and writings related to various situations
4	Community Reading Park	The Community Reading Park (Taman Bacaan Masyarakat/TBM) is a cultural institution that encourages the community's reading habits by providing and offering services in the field of reading materials, such as: books, magazines, tabloids, newspapers, comics, and other multimedia materials. It is equipped with a room for reading, discussion, book review, writing, and other literacy activities, and supported by a manager who acts as a motivator

providing internet facilities as a source of knowledge, establishing a community reading park (Taman Bacaan Masyarakat/TBM), and providing guidance to illiterate communities.

9.4 Operational Definitions for Quality Village Education

Key words related to Quality Village Education are presented in Table 9.1.

References

BPS. 2020a. *Literacy Rate of Population Aged 15 Years and Over by Place of Residence.* Available at: https://www.bps.go.id/indicator/indicator/view_data/0000/data/1461/sdgs_4/1

PISA. 2018. *Insights and Interpretations.* Available at: https://www.oecd.org/pisa/PISA%202018%20Insights%20and%20Interpretations%20FINAL%20PDF.pdf

U.S. News and World Report. 2020. *Best Countries for Education.* Available at: https://www.usnews.com/news/best-countries/best-countries-for-education

Chapter 10
SDGs Desa 5: Involvement of Village Women

10.1 Concept of Involvement of Village Women

The SDGs Desa Goal 5 aims to create conditions that place all villagers in an equitable position, without discrimination against women in all aspects of life. Fair treatment of women includes affirmative action for women to gain space in the public sphere.

Women's involvement can be achieved by encouraging and giving women confidence to actively participate in the political process and in decision-making. In the village, women must be involved in village development planning, both as participants in the village deliberation forum and as village apparatus. Women can also be involved as managers of Village-owned Enterprises (Badan Usaha Milik Desa/ BUM Desa) and participate in Cash-for-Works activities (Padat Karya Tunai Desa/ PKTD).

Over the past few decades, there has been positive development in the involvement of village women, marked by an increasing number of girls attending school, fewer girls being forced into early marriage, more women participating in the public activities, many women gaining trust as leaders, and, most importantly, in policy reform as an affirmative step towards gender justice.

Despite these achievements, there are still many challenges that need to be addressed, such as women still being less involved in activities in the village, the continued existence of physical and verbal violence against women, and so on. Therefore, the indicators of achieving the SDGs Desa 5 should also include the elimination of all forms of violence against women and children.

10.2 Targets for Involvement of Village Women

To achieve SDGs Desa 5: Involvement of Village Women, the following 13 targets have been set to be achieved by 2030:

A. H. Iskandar, *The Methods of Measuring the National Sustainable Development Goals in Indonesia*, Engaging Indonesia, https://doi.org/10.1007/978-981-96-0927-7_10

1. Gender-responsive policies that support women's empowerment are at least 30% of the total number of village regulations
2. There are regulations that guarantee women's access to services, information, and education related to family planning and reproductive health
3. The prevalence of cases of violence against female children reaches 0%
4. Cases of violence against women receiving comprehensive care services reaches 100%
5. The median age of first marriage for women (maturation of first marriage age) is above 18 years
6. The birth rate for teenage girls aged 15–19 years (Age Specific Fertility Rate/ASFR) reaches 0%
7. The Gross Enrolment Rate (GER) for senior high school/senior vocational high school/equivalent for female teenagers reaches 100%
8. The percentage of women in the Village Consultative Body (BPD) is at least 30%
9. The percentage of women as village officials is at least 30%
10. The percentage of women attending the village deliberation forum is at least 30%
11. The percentage of women participating in village development activities is at least 30%
12. Unmet need for family planning needs reaches 0%
13. Couples of Childbearing Age (Pasangan Usia Subur/PUS) understand at least four types of modern contraception methods.

10.3 Measurement and Analysis of Involvement of Village Women

10.3.1 Gender-Responsive Policies Support Women's Empowerment

Village regulations/head of village regulations/decrees related to gender and women are required in order to assess the presence of gender-responsive policies in supporting women's empowerment.

The number of these regulations is divided by the total number of existing regulations, then multiplied by 100%. If the result is more than 30%, then this SDGs Desa goal is achieved. This measurement uses the following formula:

$$\%\,Gender - Responsive\,Policies = \frac{\sum \text{Village Regulations/Head of Village Regulations/Decrees related to gender and women}}{\sum \text{Village Regulations/Head of Village Regulations/Decrees}} \times 100\%$$

The target of gender-responsive policy supports the empowerment of women, with a minimum target of 30%. The village indicators used are village regulations and village head decisions that accommodate the role of women and the number of village programmes/activities with direct female beneficiaries. The Neighbourhood Units/hamlets (Rukun Tetangga/RT) indicators used are the number of hamlet (Rukun Tetangga/RT) programmes/activities with direct female beneficiaries. The individual indicators used are female family members involved in the village deliberation forum and female family members involved in development activities. Activities that can be carried out include drafting village regulations/village head decrees on women's empowerment, implementing household-based women's economic empowerment programmes, providing business capital assistance for women, organising independent entrepreneurship training for women, forming and providing training for village cadres on gender, and providing gender-responsive planning and budgeting training to village facilitators.

10.3.2 Regulations Ensuring Women Access to Services, Information, and Education Related to Family Planning and Reproductive Health

To measure this target, data on village regulations, village head regulations or decrees that ensure women to access services, information, and education related to family planning and reproductive health are needed. If there are regulations related to this, then this target is achieved. The following formula is used to measure it:

$$\textit{Regulations related to family planning and reproductive health} = \sum \text{Family planning and reproductive health regulations}$$

This target aims for regulations that ensure women have access to services, information, and education related to family planning and reproductive health. The village indicator used is the village regulation or village head's decree on the guarantee of family planning (FP) and reproductive health services for women. Activities that can be done to achieve this target, for example, providing FP and reproductive health education to women, can improve women's welfare.

10.3.3 Prevalence of Violence Against Women and Children

Measuring the prevalence of cases of violence against children can be done by adding up cases of violence against children, dividing by the number of cases of violence that occurred in the village over the past year, then multiplying by 100%. In addition, cases of violence against women can be added up, divided by the number of cases of violence that occurred in the village over the past year, then multiplied by 100%. If the result is 0%, then the SDGs Desa target is achieved. The following formulas are used to measure this:

$$\%\ \mathit{Violence\ against\ Children} = \frac{\sum \text{Cases of violence against children}}{\sum \text{Cases of violence}} \times 100\%$$

$$\%\ \mathit{Violence\ against\ Women} = \frac{\sum \text{Cases of violence against women}}{\sum \text{Cases of violence}} \times 100\%$$

The target prevalence of cases of violence against girls is set at 0%. The individual indicators used are the number of female family members who have experienced domestic violence (DV) in the past year and the previous year, as well as the number of children who have experienced domestic violence (DV) in the past year and the previous year. Activities that can be carried out include: drafting village regulations on accompaniment and protection for victims of violence against children and women, conducting socialisation about early protection from violence against women and children, facilitating reports of sexual violence, human trafficking , and violence against women, as well as conducting training and CIE (Communication, Information and Education) on prevention and handling of violence against women and children, including human trafficking.

10.3.4 Cases of Violence Against Women who Receive Comprehensive Services

Measuring the prevalence of cases of violence against women who receive comprehensive care services can be done by adding up cases of violence against women who receive comprehensive services with the number of cases of violence against women in the village over the past year, then multiplying by 100%. If the result reaches 100%, then the SDGs Desa target is achieved. The measurement of this target uses the following formula:

$$\%\ \mathit{Violence\ against\ women\ receiving\ comprehensive\ services}$$
$$= \frac{\sum \text{Violence against women receiving comprehensive services}}{\sum \text{Cases of violence}} \times 100\%$$

The target for cases of violence against women receiving comprehensive care is set at 100%. The village indicators used are the presence of institutions dealing with violence against women, the number of reported cases of violence against women, and the number of cases of violence against women handled by the village. The individual indicators used are the provision of comprehensive care for complaints of violence against women. Actions that can be carried out include: establishing institutions/complaint posts for violence against children and women, facilitating and providing assistance to victims to get further services for acts of violence, facilitating protection of female victims of violence, and conducting socialisation about protection from violence against women and children.

10.3.5 First Marriage Age for Women

To measure the median first marriage age for women sorted the first marriage age for women from smallest to largest. If the amount of data is odd, then take the middle number. If the amount of data for the first marriage age for women is even, then add the two middle numbers, then divide by two. This value is what is called the median or middle value. If the middle value is above 18 years, then the SDGs Desa target has been achieved.

The target median first marriage age for women or maturing first marriage age is set at above 18 years. The individual indicator used is the age (date of birth) of female family members and the date of the first marriage of female family members. Activities that can be carried out include: drafting village regulations (village regulations) about the age limit for marriage in order to prevent early marriage, providing education to teenagers (15–19 years) about reproduction and family, strengthening the role of youth groups to campaign about reproductive health, optimising the role of youth groups to facilitate youth activities in the village, organising campaigns and promoting children's rights, child protection, and prevention of child marriage, as well as organising training and IEC (Communication, Information and Education) about preventing child marriage.

10.3.6 Age Specific Fertility Rate for Female Teenagers Aged 15–19 Years

To measure the ASFR, or the number of births in the group of female teenagers aged 15–19 years, the number of births in teenagers aged 15–19 years is divided by the number of teenagers aged 15–19 years, then multiplied by 100%. If the result is 0%, then the SDGs Desa target is achieved. This measurement uses the following formula:

$$\text{ASFR} = \frac{\sum \text{Birth Rate for Female Teenagers}}{\sum \text{Teenagers}} \times 100\%$$

The target for the birth rate among female teenagers aged 15–19 years (Age Specific Fertility Rate/ASFR) is set at 0%. The individual indicator used is the number of females aged 15–19 years and the number of children born to females aged 15–19 years. Activities that can be carried out include: organising reproductive health and family planning campaigns, organising campaigns against underage marriage, and formulating village regulations that regulate the age of child marriage.

10.3.7 Gross Enrolment Rate (GER) of Senior High School for Female Teenagers

To measure the Gross Enrolment Rate (GER) of senior high school/senior vocational high school/equivalent for female teenagers, data is needed on the number of female teenagers who, in a certain year and of various ages are attending high school/ vocational school/madrasah aliyah/equivalent. This number is divided by the number of female teenagers who in a certain year are in the age group 16–18 years, and then multiplied by 100%. If the result is 100%, then the SDGs Desa target is achieved. The formula is as follows:

$$\%\,GER\,of\,High\,School\,or\,equivalent = \frac{\sum \text{Female high school or equivalent students}}{\sum \text{Teenage girls aged } 16-18 \text{ years}} \times 100\%$$

The target Gross Enrolment Rate (GER) of senior high school/ senior vocational high school/equivalent is set at 100%. The individual indicators used are the age (date of birth) of female family members and the number of girls who are currently in senior high school/senior vocational high school/equivalent. Activities that can be carried out include facilitating teenage girls (15–19 years) who are not in school to participate in activities at the Community Learning Activity Centre (PKBM) in the village.

10.3.8 Women in the Village Consultative Body

To measure the percentage of women in the Village Consultative Body (BPD) the number female Village Consultative Body members is divided by the number of Village Consultative Body members, then multiplied by 100%. If the result is more than 30%, then this SDGs Desa target is achieved. This measurement uses the following formula:

$$\%\,Women\,in\,VillageConsultativeBody = \frac{\sum Women\,members\,of\,VillageCOnsultativeBody}{\sum Members\,of\,VillageConsultativeBody} \times 100\%$$

The target percentage of women in the Village Consultative Body is set at a minimum of 30%. The village indicator used is the number of women members of the Village Consultative Body. Activities that can be carried out include providing space for women's participation in the Village Consultative Body, as well as organising women's leadership training.

10.3.9 Women as Village Officials

To measure the percentage of women as village officials the number of female village officials is divided by the number of village officials, then multiplied by 100%. If the result is more than 30%, then this SDGs Desa target is achieved. This measurement uses the following formula:

$$\%\,\textit{Women as Village Officials} = \frac{\sum \text{Women as village officials}}{\sum \text{Village officials}} \times 100\%$$

The target percentage of women as village officials is set at a minimum of 30%. The village indicators used include the number of village officials, the gender of the village head, and the number of women village officials. Activities that can be carried out include providing opportunities for women's participation in village governance, as well as organizing leadership training for women.

10.3.10 Women Attending the Village Deliberation Forum

To measure the percentage of women attending the village deliberation forum (Musyawarah Desa/Musdes), the number of women attending the village deliberation forum is divided by the number of participants attending the village deliberation forum, then multiplied by 100%. If the result is more than 30%, then this SDGs Desa target is achieved. This measurement uses the following formula:

$$\%\,\textit{Women attending Village Deliberation Forum} = \frac{\sum \text{Women attending village deliberation forum}}{\sum \text{Village deliberation forum participants}} \times 100\%$$

The target percentage of women attending village deliberation forum is set at a minimum of 30%. The village indicators used are the number of village deliberation forums, the number of village deliberation forum participants, and the number of women participating in the village deliberation forum. Actions that can be carried out include, for example, providing quotas for women to be involved in village deliberation forum.

10.3.11 Women Participating in Village Development Activities

To measure the percentage of women participating in village development activities, the number of women residents participating in village development activities is divided by the number of residents participating in village development activities, then multiplied by 100%. If the result is more than 30%, then this SDGs Desa target is achieved. This measurement uses the following formula:

$$\%\ \textit{Women participating in village development activities} = \frac{\sum \text{Women participating in village development activities}}{\sum \text{Villagers participating in village development activities}} \times 100\%$$

The target percentage of women participating in village development activities is set at a minimum of 30%. The village indicators used are the number of village development activities, the number of participants/workers in village development activities, and the number of women participants/workers in village development activities. Activities that can be carried out include, for example, strengthening the role of Family Welfare Empowerment (PKK) in soliciting proposals and advocacy in the formulation of policies in the village.

10.3.12 Unmet Need for Family Planning Reaches 0%

To measure the percentage of unmet need for family planning (Keluarga Berencana/KB), the number of women who do not have access to family planning facilities is divided by the total number of women who need access to family planning facilities, then multiplied by 100%. If the result is 0%, then this SDGs Desa target is achieved. The formula used is as follows:

$$\%\ \textit{Unmet need for Family Planning (KB)} = \frac{\sum \text{Women not accessing family planning services through village facilities and outside the village}}{\sum \text{Women accessing family planning services through facilities in the village and outside the village}} \times 100\%$$

The target for unmet need for family planning is set at 0%. Indonesia ranks third highest in maternal mortality rate (MMR) compared to ASEAN countries in 2017, with 177 deaths per 100,000 births (Kemenko 2020). One of the most effective ways of reducing the MMR is by encouraging participation in the Family Planning (FP) programme. However, not all women are compliant when participating in the

FP programme, especially in the use of contraceptives. Unmet need is the unfulfilled FP need. The high figure of unmet need in Indonesia contributes to the high maternal mortality rate, this is due to unwanted pregnancies. The unmet need rate has decreased from the Indonesian Nursing Diagnosis Standard (SDKI) 1991 to 2017. However, in 2019 it increased again and has not yet reached the National Medium Term Development Planning (Rencana Pembangunan Jangka Menengah Nasional/RPJMN) target (12.1% of the target 9.9%). In measuring unmet need for FP needs, the individual indicators used are the number of women who participate in FP through contraceptive services in the village and the number of women who participate in FP through contraceptive services outside the village. Activities that can be carried out include collaborating with health centres to provide FP services and holding mobile Integrated Health Posts to provide FP and family health services.

10.3.13 Couples of Childbearing Age (COC) Understand Modern Contraceptive Methods

Modern contraceptive methods consist of birth control pills, injectable contraceptives, implants, condoms, IUDs, vasectomy, and tubectomy. To measure couples of childbearing age (COC) who understand modern contraceptive methods the number of modern contraceptive methods known to COC is divided by the number of COC. If the result is more than or equal to four, then this SDGs Desa target is achieved. The formula used is as follows:

$$\textit{Couple of Childbearing Age (COC) understand modern contraceptive methods}$$
$$= \frac{\sum \text{Modern contraceptive methods known by couples of childbearing age}}{\sum \text{Couples of childbearing age}}$$

The final target is for couples of childbearing (COC) age to understand at least four types of modern contraception methods. The village indicator used is the socialisation activities about contraception methods. The individual indicator used is the COC age who receive socialisation about contraception methods. Activities that can be done include: strengthening the role of Integrated Service Posts (Pos Pelayanan Terpadu/Posyandu) to monitor young couples, as well as holding consultations about contraception.

10.4 Operational Definitions for Involvement of Village Women

Key words related to Involvement of Village Women are presented in Table 10.1.

Table 10.1 Operational definitions for women's village involvement

	Key word	Operational definition
1	Maternal mortality rate (MMR)	The number of women's deaths during pregnancy or within 42 days of termination of pregnancy, regardless of the duration and place of delivery, caused by the pregnancy or its management, and not by other causes (accidents, falls, etc.) per 100,000 live births
3	*Human trafficking*	The act of recruitment, transportation, harbouring, sending, moving, or receiving a person by threat of violence, use of violence, abduction, confinement, forgery, fraud, abuse of power or a vulnerable position, debt entrapment or giving payment or benefits, so as to obtain the consent of a person having control over another person, whether carried out within the country or between countries, for the purpose of exploitation or resulting in the person being exploited
2	Family planning (FP)	A movement to form a healthy and prosperous family by limiting births
	Reproductive health	A state of complete physical, mental, and social well-being, not merely free from disease or disability, in all matters relating to the reproductive system, its functions and processes (WHO)
3	Contraception	Efforts to prevent pregnancy that are temporary or permanent which can be done without using tools, mechanically, using tools/drugs, or by surgery
4	Couples of fertile age (PUS)	Couples with a wife aged 15–49 years or a married couple aged less than 15 years and already menstruating or a wife aged 15 years but still menstruating (having periods)
5	Development	Activities/programmes aimed at improving the quality of life and living for the greatest welfare of the village community
6	Women's empowerment	Activities carried out to mobilise women to take the initiative to start social activities to improve their own situation and conditions
7	Early marriage	Marriage that takes place at an age below the reproductive age, which is less than 18 years for women and less than 25 years for men

Reference

Kemenko, P.M.K. 2020. *Reduction of unmet need rate decreases maternal mortality rate.* [internet]. https://www.kemenkopmk.go.id/penurunan-angka-unmet-need-turunkan-angka-kematian-ibu

Chapter 11
SDGs Desa 6: Village with Clean Water and Sanitation

11.1 Concept of Village with Clean Water and Sanitation

Clean water and adequate sanitation are basic human needs. Fulfilling clean water and adequate sanitation is still a problem worldwide. Therefore, ensuring village communities have access to clean water, drinking water, and sanitation is the goal of SDGs Desa.

Drinking water that is said to be decent is household drinking water that goes through a treatment process, or without a treatment process yet meets health requirements and can be drunk directly, while sanitation is said to be decent if it meets health requirements, that is, it is used by the household alone or shared with certain other households and is equipped with a swan neck type toilet, and the final disposal place for faeces is a septic tank (Kemen PUPR n.d.).

In the period from 2015 to 2020, the percentage of household access to decent drinking water, both in urban and rural areas, increased, as shown in Fig. 11.1.

Similarly to the data on household access to clean and adequate drinking water, the percentage of household access to adequate sanitation has increased. Both in urban and rural areas, household access to adequate sanitation from 2015 to 2020 increased, as shown in Fig. 11.2.

The achievement of this SDGs Desa can be measured by several things, such as household access to drinking water and adequate sanitation reaching 100% by 2030, the occurrence of drinking water use efficiency, and the presence of actions to protect and restore ecosystems related to water resources, including mountains, forests, wetlands, rivers, groundwater, and lakes.

A. H. Iskandar, *The Methods of Measuring the National Sustainable Development Goals in Indonesia*, Engaging Indonesia, https://doi.org/10.1007/978-981-96-0927-7_11

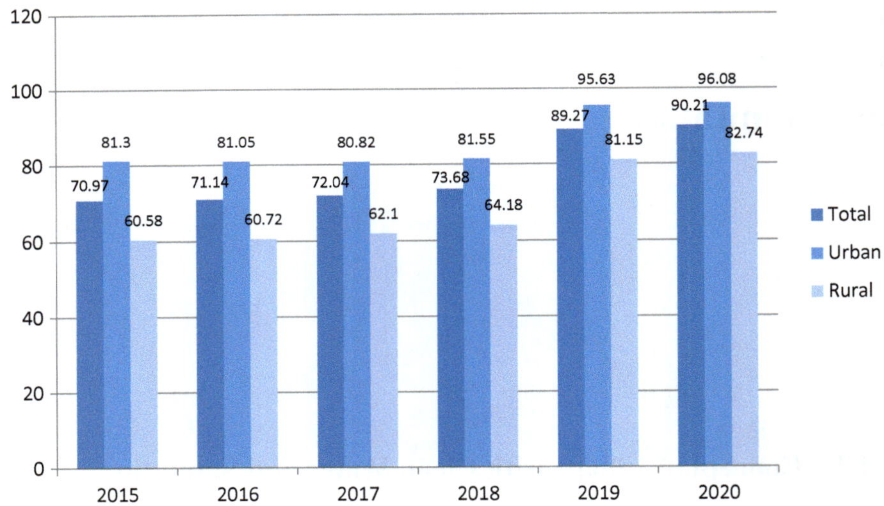

Fig. 11.1 Percentage of household access to decent drinking water, 2015–2020

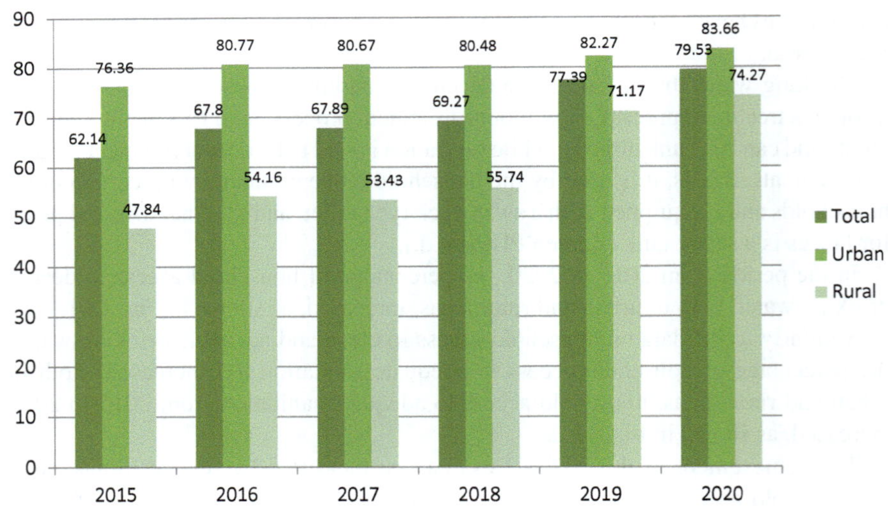

Fig. 11.2 Percentage of household access to adequate sanitation, 2015–2020

11.2 Village with Clean Water and Sanitation Targets

For SDGs Desa 6: Village with Clean Water and Sanitation, the following 22 targets have been set to be achieved by 2030:

1. Access to adequate drinking water services reaches 100% of families
2. Families and industries served by raw water reach 100%
3. Family access to adequate sanitation reaches 100%
4. Families implementing Community-Based Total Sanitation (STBM) reach 100%
5. Families and industries using wastewater facilities reach 100%
6. Families and industries using faecal sludge management facilities reach 100%
7. The quality of water sources in Neighbourhood Units/hamlets (Rukun Tetangga/RT) is colourless, odourless, tasteless
8. There are regulations on the use of groundwater
9. Reduction of irrigation water costs in industries for those who apply safe wastewater for agriculture
10. Availability of environmental conservation regulations around river flows
11. Availability of information from the nearest hydrological and climatological station
12. Availability of water resource information
13. The number of springs in the Neighbourhood Units/hamlet (Rukun Tetangga/RT) remains
14. There are tree planting activities around river flows
15. There are communal activities for the preservation of river and lake flows
16. There are regulations on water resource management
17. Implementation of training/socialisation of water resource management
18. The quality of lake water is colourless, odourless, tasteless
19. Implementation of annual lake-dredging activities
20. No annual lake erosion events
21. No barren/bald land
22. The number of springs in the village environment remains.

11.3 Measurement and Analysis of Village with Clean Water and Sanitation

11.3.1 Access to Decent Drinking Water Services

To measure access to drinking water services, the number of families who have a source of drinking water is divided by the number of families in the village, then multiplied by 100%. If the result is 100%, then this SDGs Desa target is achieved.

The formula used is as follows:

$$\textit{Decent drinking water (RT)} = \sum \textit{Infrastructure of decent drinking water}$$

$$\textit{Decent drinking water (Families)} = \frac{\sum \text{Families having source of drinking water}}{\sum \text{Families}} \times 100\%$$

The target for access to potable water services is set at 100% of families. The Neighbourhood Units/hamlets (Rukun Tetangga/RT) indicator used is the existence of a suitable drinking water infrastructure. The individuals used are the use of drinking water infrastructure (distance of the house from the drinking water source) and water quality (taste, smell, and colour). Activities that can be carried out include building drinking water/clean water facilities managed by the Village-owned Enterprises (Badan Usaha Milik Desa/BUM Desa) for the community, through procurement of pumps, construction of water storage tanks, construction of drinking-water/clean-water pipelines to residents' homes, procurement of water tanks for households. Other activities include the procurement of water tank trucks in collaboration with other villages for the distribution of drinking water/clean water for households that are not near to water pipes, conducting reforestation in the water source area, providing training in drinking water management, and building infiltration wells.

11.3.2 Families and Industries Served by Raw Water

To measure the families served by raw water, the number of families served by raw water is divided by the number of families in the village, and then multiplied by 100%. If the result is 100%, then this SDGs Desa target is achieved. The formula used is as follows:

$$\textit{Families served with Raw Water} = \frac{\sum \text{Families served with raw water}}{\sum \text{Families}} \times 100\%$$

To measure the industry served by raw water, the number of industries served by raw water is divided by the number of industries in the village, then multiplied by 100%. If the result is 100%, then the SDGs Desa target is achieved. The formula used is as follows:

$$\textit{Industry served with Raw Water} = \frac{\sum \text{Industry served with raw water}}{\sum \text{Industries}} \times 100\%$$

The target for families and industries served by raw water is set at 100%. The Neighbourhood Units/hamlets (Rukun Tetangga/RT) indicator used is the number of raw water infrastructure sources (springs, groundwater, surface water, and rainwater) and the use of raw water for industry. Meanwhile, the family indicator used is the use of springs, groundwater, surface water, and rainwater as drinking water. Activities that can be carried out include, for example, providing raw water managed by Village-owned Enterprises (BUM Desa).

11.3.3 Family Access to Adequate Sanitation

To measure access to adequate sanitation, data on the number of families that have communal facilities for bathing, washing, and urinating/defecating (Mandi, Cuci, Kakus/MCK). This number is divided by the number of families in the village, then multiplied by 100%. If the result is 100%, then the SDGs Desa target is achieved. The formula used is as follows:

$$\textit{Adequate sanitation} = \frac{\sum \text{Families having communal facilities for bathing, washing, and urinating/defecating (MCK)}}{\sum \text{Families}} \times 100\%$$

The target for family access to adequate sanitation is set at 100%. The family indicators used are bathing, washing, and urinating/defecating (MCK) facilities (private/group/public/none), bathing water source (piped/river or lake/rainwater catchment/other), defecation facilities (private/shared/public/other), and liquid waste disposal sites (tank/field or river or sea/hole in the ground/other). Activities that can be carried out include: creating or rehabilitating public facilities for bathing, washing, and urinating/defecating (MCK) for adults and children; building communal household waste channels; conducting training for community health cadres related to sanitation; conducting campaigns and socialisation related to sanitation; organising campaigns and socialisation in the community on use of facilities for bathing, washing, and urinating/defecating (MCK); conducting socialisation about healthy sanitation standards; drafting village regulations about the obligation to build healthy household sanitation; building residential drains; organising a Free Open Defecation Village campaign; and building a household-waste-management system.

11.3.4 Families Implementing Community-Led Total Sanitation

To measure families implementing Community-Led Total Sanitation (Sanitasi Total Berbasis Masyarakat/STBM) data is needed on the number of families implementing Community-Led Total Sanitation, and this number is divided by the number of families in the village, then multiplied by 100%. If the result is 100%, then the SDGs Desa target is achieved. The formula used is as follows:

$$\textit{Community} - \textit{Led Total Sanitation} = \frac{\sum \text{Families implementing Community} - \text{Led Total Sanitation}}{\sum \text{Families}} \times 100\%$$

The target for families implementing Community-Led Total Sanitation is set at 100%. The family indicators used are hand-washing facilities, defecation facilities, household drinking water and food management infrastructure, household waste security infrastructure, and household liquid waste security infrastructure. Activities that can be carried out include building waste disposal sites, providing waste carts and waste-transport vehicles, compiling waste-management and processing programmes in villages, providing socialisation and facilitating the construction of Community-Led Total Sanitation, providing community environmental health services, and conducting waste-recycling training.

11.3.5 Families and Industries Using Wastewater Facilities

If the village has wastewater facilities, then the target families and industries using wastewater facilities can be measured. To measure the families using wastewater facilities, the number of families using wastewater facilities is divided by the number of families in the village, then multiplied by 100%. If the result is 100%, then the SDGs Desa target is achieved. The formula is as follows:

$$\%\,\textit{Families using Wastewater Facilities} = \frac{\sum \text{Families using wastewater facilities}}{\sum \text{Facilities}} \times 100\%$$

To measure the industry using wastewater facilities, the number of industries using wastewater facilities is divided by the number of industries in the village, then multiplied by 100%. If the result is 100%, then the SDGs Desa target is achieved. The formula is as follows:

$$\%\,\textit{Families using Wastewater Facilities} = \frac{\sum \text{Industries using wastewater facilities}}{\sum \text{Facilities}} \times 100\%$$

The target for families and industries using wastewater facilities is set at 100%. The village indicator used is the number of wastewater infrastructure on a settlement scale. The family indicator used is the family wastewater disposal facility. Activities that can be carried out include building waste-management sites, building sanitation systems for managing settlement wastewater, procuring waste-processing machines, and conducting household waste and business-waste processing training.

11.3.6 *Families and Industries Using Sludge Management Facilities*

If the village has sludge management facilities, then the target for families and industries using wastewater facilities can be measured. To measure the families using sludge management facilities, the number of families using sludge management facilities is divided by the number of families in the village, then multiplied by 100%. If the result is 100%, then the SDGs Desa target is achieved. The formula is as follows:

$$\% \, Families \, using \, Sludge \, Management \, Facilities = \frac{\sum \text{Families using sludge management facilities}}{\sum \text{Families}} \times 100\%$$

To measure industries using sludge management facilities, the number of industries using sludge management facilities is divided by the total number of industries in the village, and then multiplied by 100%. If the result is 100%, then the SDGs Desa target is achieved. The formula is as follows:

$$\% \, Industries \, using \, Sludge \, Management \, Facilities = \frac{\sum \text{Industries using sludge management facilities}}{\sum \text{Industries}} \times 100\%$$

The targets for families and industries using sludge management facilities are set at 100%. The village indicator used is the Sludge Treatment Plant (STP) management facility. The family indicator used is the use of STP management facilities. Activities that can be carried out include providing community-based STP facilities and improving village sanitation services. For example, building STP management facilities, building public toilets in settlements, carrying out sanitation programmes, and procuring and maintaining facilities for. for bathing, washing, and urinating/defecating.

11.3.7 *Quality of Water Sources in Neighbourhoods is Colourless, Odourless, Tasteless*

To measure the quality of colourless, odourless, and tasteless water sources throughout the Neighbourhood Units/hamlets (Rukun Tetangga/RT), the number of unpolluted water sources in each Neighbourhood Units/hamlets (Rukun Tetangga/RT) is divided by the number of water sources in each Neighbourhood Units/hamlets (Rukun Tetangga/RT), and then multiplied by 100%. If the result is 100%, then the SDGs Desa target is achieved. The formula is as follows:

$$Good\ Quality\ of\ Water\ Sources = \frac{\sum \text{Unpolluted water sources in hamlet (RT)}}{\sum \text{Water sources}} \times 100\%$$

The target quality of water sources in the Neighbourhood Units/hamlets (Rukun Tetangga/RT) is aimed to be colourless, odourless, and tasteless. The Neighbourhood Units/hamlets (Rukun Tetangga/RT) indicator used is the quality of water sources. Activities that can be carried out include cooperating with the Public Works Department to check the quality of water sources, developing a program on improving the quality of water sources with reboisation or other programs, as well as greening in the watershed area.

11.3.8 Groundwater Use Regulation

To measure the existence of regulations on the use of groundwater a Village Regulation, Village Head Regulation or decrees on the use of groundwater are needed. If there are regulations related to this matter, then the target is achieved. The following formula is used to measure it:

$$Groundwater\ Use\ Regulation = \sum \text{Groundwater use regulation}$$

This target aims for the existence of groundwater use regulations. The village indicators used are the existence of Village Regulations/Village Head Decrees on good water usage and the existence of Village-owned Enterprises (BUM Desa) that manage raw water economic area. Activities that can be carried out include building/improving village-scale reservoirs, greening in the area Watershed (Daerah Aliran Sungai/DAS), compiling Village Regulations/Village Head Decrees on management and use of raw water, managing raw water through Village-owned Enterprises (BUM Desa), and arranging distribution of village raw water for economic activities.

11.3.9 Irrigation Water Costs in Industry for Those Implementing Safe Wastewater for Agriculture

To measure the cost of irrigation water in the industry for those implementing safe wastewater for agriculture requires the total cost of agricultural irrigation use in a particular year, reduced by the total cost of agricultural irrigation use in the previous year, divided by the total cost of agricultural irrigation use in the previous year, then multiplied by 100%. If the result is negative (indicating a reduction), then it has reached the SDGs Desa target. The formula is as follows:

$$\textit{Agricultural Irrigation Water Costs} = \frac{\sum \text{Cost of agricultural irrigation use in a certain year} - \sum \text{Cost of agricultural irrigation use in the previous year}}{\sum \text{Cost of agricultural irrigation use in the previous year}} \times 100\%$$

This target aims at reducing the cost of irrigation water in industries that implement safe wastewater for agriculture. The village indicator used is the water-saving incentive for agriculture in industries that process wastewater. While the family indicator used is the cost of agricultural irrigation use over the past year and the year before. Activities that can be carried out include, compiling Village Regulations on the incentive for water use for agriculture, collaborating with the Agriculture Department to provide counselling and assistance in saving water use for economic activities, as well as building a village irrigation system through the Cash-for-Work (Padat Karya Tunai Desa/PKTD) scheme.

11.3.10 Regulation of Environmental Preservation Around River Streams

To measure the existence of regulations on environmental preservation around river streams, Village Regulations, Village Head Regulations, and the Village Head Decree on environmental preservation around river streams are needed. If there are regulations related to this, then the target is achieved. To measure it, the following formula is used:

$$\textit{Regulations of Environmental Preservation around River Streams} = \sum \text{Regulations of environmental preservation around river streams}$$

This target aims for the existence of environmental preservation regulations around river flows. The village indicator used is the existence of Village Regulations/Head of Village Decree on environmental preservation around river flows. Activities that can be carried out include, compiling a *master plan* for the village on planning and managing river catchment areas, and collaborating with the Public Works Department to synchronise and harmonise the planning of River Catchment Areas (RCA).

11.3.11 Information from the Nearest Hydrology and Climatology Station

To measure this target, if the village has information from the nearest hydrology and climatology station then this target is achieved. The formula used is as follows:

$$Information\ from\ the\ Nearest\ Hydrology\ and\ Climatology\ Station$$
$$= \sum \text{Information from the nearest hydrology and climatology station}$$

This target aims for the availability of information from the nearest hydrology and climatology station. The village indicator used is the availability of information from the nearest hydrology and climatology station. Activities that can be carried out include collecting data on water and climate conditions in the village, compiling a data bank on current hydrology and climatology, and creating an environmental information board.

11.3.12 Water Resource Information

To measure this target, if the village has information related to water resources then this target is achieved. The formula used is as follows:

$$Water\ Resources\ Information = \sum Water\ resources\ information$$

This target aims for the availability of water resource information. The village indicator used is the availability of water resource information. Activities that can be carried out include collecting data on water resources in the village, periodically updating water resource data, and mapping potential village water resources to support community needs.

11.3.13 Number of Springs in the Neighbourhood Units

To measure the number of springs in the Neighbourhood Units/hamlets (Rukun Tetangga/RT) environment, the number of springs in the Neighbourhood Units/ hamlets (Rukun Tetangga/RT) environment in a certain year is needed, and then the number of springs in the Neighbourhood Units/hamlets (Rukun Tetangga/RT) environment in the previous year is subtracted from this. The result is divided by the number of springs in the Neighbourhood Units/hamlets (Rukun Tetangga/RT) environment in the previous year, then multiplied by 100%. If the result is greater than or equal to 0% (indicating an increase or steady), then it has reached the SDGs Desa target. The formula is as follows:

$$\%\,\text{Number of Springs in the Neighborhood Units (RT)}$$
$$= \frac{\sum \textit{Spring}\ \text{sin}\ \textit{RTs in a Certain year} - \sum \textit{Spring}\ \text{sin}\ \textit{RTs in the previous year}}{\sum \textit{Spring}\ \text{sin}\ \textit{RTs in the previous year}} \times 100\%$$

This target aims to maintain the number of springs in the Neighbourhood Units / hamlets (Rukun Tetangga/RT). The Neighbourhood Units/hamlets (Rukun Tetangga/ RT) indicator used is the number of springs over the past year and the number of springs in the previous year in the Neighbourhood Units/hamlets (Rukun Tetangga/ RT). Activities that can be carried out include reforestation around the watershed area, establishing posts/task forces responsible for protecting and managing water resources, and fostering cooperation between areas in maintaining and managing the watershed.

11.3.14 Tree-Planting Activities Around the River Flow

To measure the presence of tree-planting activities in critical lands, such as around river flows and dredging of rivers and lakes, data on tree-planting activities in critical lands and the involvement of villagers in these activities (all involved or partially involved) are needed. If there are tree-planting activities on critical land in the village, then the SDGs Desa target is achieved. The formula used is as follows:

$$\textit{Tree Planting Activities around the River Flow}$$
$$= \sum \text{Tree planting activities around the river flow}$$

This target aims for tree-planting activities around river streams. The village indicator used is the number of tree-planting activities around river streams in the past year. Activities that can be carried out include compiling a planting program around the watershed and establishing village forestry areas around the river stream area.

11.3.15 Community Service Activities for River and Lake Conservation

To measure this target, information is needed on the number of community service activities to conserve river and lake flows, and the involvement of village residents in these activities (all involved or partially involved). If there are community service activities to conserve river and lake flows in the village, then the SDGs Desa target is achieved. The formula used is as follows:

$$\textit{Community Service Activities for River and Lake Conservation} = \sum \text{Community Service activities for river and lake conservation}$$

This target aims for the existence of community services activities for river and lake conservation. The village indicator used is the number of community conservation activities to preserve river and lake flows. The family indicator used is participation in community conservation activities to preserve river and lake flows. Activities that can be carried out include utilising lakes and rivers as tourism facilities, giving authority to Village-owned Enterprises (BUM Desa) and community groups to manage tourism in rivers or lakes, and creating fish ponds as a place for freshwater fish cultivation.

11.3.16 Water Resource Management Regulation

To measure the existence of regulations on water resource management, village regulations, village head regulations, and village head decision letters on water resource management are needed. If there are regulations related to this, then the target is achieved. To measure it, the following formula is used:

$$\textit{Water Resource Management Regulation} = \sum \text{Water resource management regulation}$$

This target aims for the regulation of water resources management. The village indicator used is the existence of village regulations/village head decrees on water resources management. Activities that can be carried out, for example, drafting village regulations/village head decrees on water resources management and establishing a post/working group focused on managing water resources.

11.3.17 Training/Socialisation on Water Resources Management

To measure this target, information on the number of training/socialisations on water resources management and the involvement of village residents in these activities (all involved or partially involved) is needed. If there is training/socialisation on water resources management, then the SDGs Desa target is achieved. The formula used is as follows:

$$\textit{Training or Socialisation of Water Resources Management} = \sum \text{Training or socialisation of water resources management}$$

The goal is to implement training/socialisation on water resource management. The village indicator used is the existence of water resource management training/ socialisation. Activities that can be done, for example, providing training on water resource management.

11.3.18 Lake Water Quality is Colourless, Odourless, Tasteless

To measure the quality (colourless, odourless, and tasteless) of water from unpolluted rivers/irrigation channels/lakes/reservoirs, the number of unpolluted rivers/irrigation channels/lakes/reservoirs is divided by the number of rivers/irrigation channels/lakes/ reservoirs in the village then multiplied by 100%. If the result is 100%, then the SDGs Desa target is achieved. The formula is as follows:

$$\text{The Quality of Unpolluted Rivers/Irrigation Channels/Lakes/Reservoirs} = \frac{\sum \text{Unpolluted rivers/irrigation channels/lakes/reservoirs}}{\sum \text{Rivers/irrigation channels/lakes/reservoirs}} \times 100\%$$

The target for the quality of river/irrigation channel/lake/reservoir water is that it should be colourless, odourless, and tasteless. The Neighbourhood Unit/hamlets (Rukun Tetangga/RTs) indicator used is the quality (colour, smell, and taste) of the lake water used. Activities that can be carried out include regular measurements of water quality in the village, greening the lake area, managing the lake area as a tourist area, and regular cleaning in the lake area.

11.3.19 Annual Lake-Dredging Activities

To measure this target, information on the number of annual lake-dredging activities and the involvement of village residents in these activities (all involved or some involved) is needed. If there are annual lake dredging activities, then the SDGs Desa target is achieved. The formula used is as follows:

$$\textit{Annual Lake Dredging Activities} = \sum \text{Annual lake dredging activities}$$

This target aims for annual lake-dredging activities. The village indicator used is the presence of lake-dredging activities in the past year. Activities that can be carried out include dredging/deepening the lake in the village, as well as regularly checking the lake's sustainability.

11.3.20 Annual Lake-Erosion Events

To measure this target, information on the number of lake-erosion incidents over the past year is required. If there have been no lake-erosion incidents over the past year, then the SDGs Desa target has been achieved. The formula used is as follows:

$$\textit{Annual Lake Erosion Events} = \sum \textit{Lake erosion events over the past year}$$

This target aims for no annual lake-erosion events. The village indicator used is the number of lake-erosion events over the last year. Activities that can be carried out include recording the level of erosion in the village lake, building lake embankments, and constructing erosion-retaining walls.

11.3.21 Barren/Bare Land

To measure this target, information on the amount of barren/bare land in the village is needed. If there is no barren/bare land in the village, then the SDGs Desa target is achieved. The formula used is as follows:

$$\textit{Barren or Bare Land} = \sum \textit{Barren or bare land}$$

This target aims to ensure that there is no barren/bare land in the village. The Neighbourhood Units/hamlets (Rukun Tetangga/RT) indicator used is the area of critical land (barren/bare). Activities that can be carried out include forming forestry partnerships, collaborating with the Forestry Department for planting and rehabilitating of critical land, engaging in critical forest rehabilitation with community participation, and rehabilitating mangrove areas.

11.3.22 Number of Springs in the Village Environment

To measure the number of springs in the village environment, the number of springs in the village environment in a certain year is needed, then the number of springs in the village environment in the previous year is subtracted from this. The result is divided by the number of springs in the village environment in the previous year, then multiplied by 100%. If the result is greater than or equal to 0% (indicating an increase or stability), then it has achieved the SDGs Desa target. The formula is as follows:

$$\% \text{ Number of Springs in the Village Environment} = \frac{\textit{Springs in the village environment in a certain year} - \textit{Springs in the village environment in the previous year}}{\text{Springs in the village environment in the previous year}} \times 100\%$$

This target aims to maintain the number of springs in the village environment. The village indicators used are the number of springs this year and the number of springs last year. Activities that can be carried out include compiling village regulations on spring protection, creating posts/working groups tasked with monitoring and preserving springs, and building spring ponds.

11.4 Operational Definitions for Village with Clean Water and Sanitation

Key words related to Village with Clean Water and Sanitation are presented in Table 11.1.

Table 11.1 Operational definitions for village with clean water and sanitation

	Key word	Operational definition
1	Raw water	Water that can come from surface water sources, groundwater basins and/or rainwater that meets certain quality standards as raw water for drinking water
2	Freshwater fish farming	A form of fish farming in freshwater with a salinity of less than 0.05%
3	Lake/ reservoir/ dam	Land located in a large basin on the earth's surface that is flooded by water, either fresh or salty, with the entire basin surrounded by land
4	Watershed/ riverbank	A stretch of territory/area bounded by topographic boundaries (hill ridges) that serves to receive, collect rainwater, sediment, and nutrients and channel them through tributaries and exit at one point (outlet).
5	Small reservoir	A catchment basin used to regulate and store rainwater flow and
6	Erosion	The event of erosion of solids due to transportation by wind, water, ice, rain, creep on soil and other materials
7	Septage treatment plant (STP)	A wastewater treatment plant designed to only receive septage via a vehicle (septage truck)
8	Clean water quality	A measure of water condition seen from its physical, chemical and biological characteristics
9	Critical land	Land that has experienced damage so that it loses or reduces its function
10	Spring	A natural condition where groundwater flows out from an aquifer to the ground surface
11	Sanitation	Efforts to foster and create a good condition in the field of health, especially public health
12	Water resources	Resources in the form of water that have potential for human use
13	River	A large and elongated flow of water that flows continuously from upstream to downstream

Chapter 12
SDGs Desa 7: Clean and Renewable Energy Village

12.1 Concept of Clean and Renewable Energy Village

SDGs Desa Goal 7 will ensure all villagers have access to renewable energy. The achievement of this goal by 2030 can be measured by several indicators, including household electricity consumption in the village reaching at least 1,200 kWh, families in the village using gas or wood waste for cooking, and the use of a mix of renewable energy in the village.

Clean energy is energy produced with minimal negative impact on social, cultural, health, and environmental aspects (Rohim 2009). On the other hand, renewable energy is energy derived from renewable energy sources, including geothermal heat, wind, bioenergy, sunlight, water flow and falls, and the movement and temperature differences of ocean layers (BPS n.d.).

Improving infrastructure and enhancing supporting technology to provide cleaner and more efficient energy can help reduce environmental impact (Institute for Essential Services Reform 2019). Currently, not all villagers have access to clean energy, and not all potential renewable energy in the village is optimised for meeting the energy needs of the villagers. However, the potential for renewable energy in Indonesia is abundant.

12.2 Targets for Clean and Renewable Energy Village

To achieve the goal of a Village with Clean and Renewable Energy, the following five targets must be achieved by 2030:

1. The number of families using electricity reaches 100%
2. Per capita electricity consumption is more than 1,200 kWh

A. H. Iskandar, *The Methods of Measuring the National Sustainable Development Goals in Indonesia*, Engaging Indonesia, https://doi.org/10.1007/978-981-96-0927-7_12

3. Families using gas or wood waste for cooking reaches 100%
4. The use of a mix of renewable energy reaches 60% of families
5. Reduction in families using oil for transportation and cooking reaches 50%.

12.3 Measurement and Analysis of Clean and Renewable Energy Village

12.3.1 Number of Families Using Electricity

To measure the percentage of families using electricity (National Electricity Company/PLN or non-PLN), the number of families using electricity is divided by the number of families in the village, then multiplied by 100%. If the value is 100%, then the SDGs Desa target has been achieved. The formula used is as follows:

$$\%\, Electricity\ users = \frac{\sum \text{Families using electricity}}{\sum \text{Families}} \times 100\%$$

The target number of families using electricity is set at 100%. The family indicator used is the number of families using electricity and the type of electricity source used. Activities that can be carried out include building micro hydropower plants, biodiesel, biogas, solar, wind or other primary energy working together with Village-owned Enterprise, and installing a network of non-PLN (non-National Electricity Company)electricity distribution to residents' homes.

12.3.2 Level of Electricity Consumption per Capita

To measure the level of electricity consumption per capita, data for the total National Electricity Company (PLN) and non-PLN electricity consumption is divided by the number of residents in the village. If the result is more than 1,200 KWh, then it has reached the SDGs Desa target. The formula is as follows:

$$Electricity\ Consumption\ per\ Capita = \frac{\sum \text{National Electricity Company (PLN) and non} - \text{PLN electricity consumption}}{\sum \text{Population}}$$

The target for per capita electricity consumption is set at more than 1,200 KWh. The family indicator used is the use of electricity (in KWh or IDR) during the previous month and the time (hours) of electricity use during the same month. Activities that can be done include facilitating government electricity subsidies for poor households, as well as carrying out village funded Cash-for-Work such as training for processing and marketing of village/MSMEs flagship commodities to increase the purchasing power of the community.

12.3.3 Families Using Gas or Wood Waste for Cooking

To measure the percentage of families using gas or wood waste for cooking, the number of families using gas or wood waste for cooking is divided by the total number of families, then multiplied by 100%. If the value is 100%, then the SDGs Desa target has been achieved. The formula is as follows:

$$\begin{aligned}&\%\ \mathit{Users\ of\ gas\ or\ wood\ waste\ for\ cooking}\\&=\frac{\sum \text{Families using gas or wood waste for cooking}}{\sum \text{Families}}\times 100\%\end{aligned}$$

The target for families using gas or wood waste for cooking is set at 100%. The indicators used are the number of families using gas for cooking, as well as the number of families using wood waste for cooking. Activities that can be carried out include organising campaigns for the use of non-LPG (Liquid Petroleum Gas), developing biogas, installing biogas networks to households, and facilitating the data collection of natural gas network installations to households.

12.3.4 Use of Renewable Energy Mix

To measure the percentage of families using a renewable energy mix, the number of users of the renewable energy mix is divided by the number of families in the village, then multiplied by 100%. If the value is more than 60%, then the SDGs Desa target has been achieved. The formula is as follows:

$$\%\ \mathit{Users\ of\ Renewable\ Energy\ Mix}=\frac{\sum \text{Users of Renewable Energy Mix}}{\sum \text{Families}}\times 100\%$$

The renewable energy mix target is set to reach 60% of families. The family indicator used is the number of families that use renewable energy, as well as the type of renewable energy used, for example, firewood waste, wind, micro hydro, and solar light. Activities that can be done include, for example, developing new renewable energy as an alternative source of electricity in the village. In addition to this, it is also possible to use micro hydropower plants, biodiesel, sun, and wind. This effort can be done through cooperation with the Village-owned Enterprises.

Table 12.1 Operational definitions for clean and renewable energy village

	Key word	Operational definition
1.	Biogas	Energy produced from organic waste such as animal manure, or kitchen waste like used vegetables
2.	Power plant	Part of industrial equipment used to produce and generate electricity from various energy sources

12.3.5 Reduction of Kerosene Users for Transportation and Cooking

To measure the reduction of families using kerosene or coal for transportation and cooking, the number of families who use kerosene or coal for transportation and cooking in a certain year, minus the number of families who use kerosene or coal for transportation and cooking in the previous year, is divided by the number of families who use kerosene or coal for transportation and cooking in the previous year, then multiplied by 100%. If the value reaches −50% (depicting a reduction), then the SDGs Desa target has been achieved. The formula is as follows:

$$\%\ \text{Reduction of kerosene or coal users for transportation and cooking}$$
$$= \frac{\begin{array}{l}\text{Users of kerosene or coal for transportation and cooking in a certain year} - \\ \text{Users of kerosene or coal for transportation and cooking in the previous year}\end{array}}{\text{Users of kerosene or coal for transportation and cooking in the previous year}} \times 100\%$$

The target for reducing families using oil for transportation and cooking is set to reach 50%. The family indicator used is the use of electric/gas/oil/other energy (in IDR) for cooking as well as the use of electric/gas/oil/other energy (in IDR) for transportation. Activities that can be done include installing non-PLN (National Electricity Company) electric power distribution networks to residents' homes, as well as installing biogas installations for poor households.

12.4 Operational Definitions for Clean and Renewable Energy Village

Key words related to the Clean and Renewable Energy Village are presented in Table 12.1.

References

BPS. N.d. Renewable energy mix. [internet]. https://sirusa.bps.go.id/sirusa/index.php/indikator/1563.

Institute for Essential Services Reform. 2019. Sustainable energy access for rural communities: Status, challenges, and opportunities. [internet]. http://iesr.or.id/wp-content/uploads/2019/05/Proceeding-PE-11.pdf.

Rohim M. 2009. *Community Guide for Environmental Health*. Bandung (ID): Tambuhak Sinta Foundation.

Chapter 13
SDGs Desa 8: Equitable Village Economic Growth

13.1 Concept of Equitable Village Economic Growth

The goal of SDGs Desa 8 is to realise equitable village economic growth. The World Development Report 2006 (World Bank 2005) explains that equality is defined in two basic principles. The first is that opportunity is the same for all; so a person's life achievements should, first and foremost, be determined by their talents and efforts, and not by various circumstances that are their fate, such as race, gender, social and family background, or country of birth. The second principle is the avoidance of harmful results, especially in terms of health, education, and consumption.

The main target of the SDGs Desa is economic growth, accompanied by the equitable distribution of development results, through creating decent jobs, and opening up new economic opportunities for all village residents. The success indicators of this goal include: the absorption of the workforce in jobs; the implementation of Cash-for-Work (Padat Karya Tunai Desa/PKTD)[1] programmes capable of absorbing 50% of the village workforce; and workplaces that provide a sense of security and are equipped with health service facilities.

13.2 Targets for Equitable Village Economic Growth

To achieve the goal of Equitable Village Economic Growth by 2030, the following 13 targets have been set:

1. Average per capita Village GDP above IDR 30 million
2. Average per capita Village GDP above IDR 30 million per worker in the village
3. The percentage of workers in the formal sector reaches 51%
4. At least 51% of micro, small and medium enterprises (MSMEs) in the village have access to formal capital
5. The number of people working reaches 100% of the workforce

A. H. Iskandar, *The Methods of Measuring the National Sustainable Development Goals in Indonesia*, Engaging Indonesia, https://doi.org/10.1007/978-981-96-0927-7_13

6. The level of open unemployment in the village reaches 0%
7. The absorption of labour in Cash-for-Work Programme (Padat Karya Tunai Desa/PKTD) reaches more than 50% of the unemployed in the village
8. New workforce who have never worked receive 100% skills training
9. Acts of violent abuse, exploitation, neglect, and other mistreatment of children reach 0%
10. Workplaces that have health and safety facilities reach 100%
11. The contribution of tourism activities reaches at least 8% of the village GDP
12. The number of tourists increases every year
13. There is access to capital and financial services.

13.3 Measurement and Analysis of Equitable Village Economic Growth

13.3.1 Village GDP per Capita

To measure the village GDP per capita, the village GDP is needed, which is the total of village funds: Village Own-Source Revenue (Pendapatan Asli Desa/PADes), local tax revenues, Village Fund Allocation (Alokasi Dana Desa/ADD), financial assistance from the Regional Revenue and Expenditure Budget (Anggaran Pendapatan dan Belanja Daerah/APBD) of the province, financial assistance from the Regional Revenue and Expenditure Budget (Anggaran Pendapatan dan Belanja Daerah/APBD) of the regency/city, grants, and others. This figure is then divided by the number of village residents. If the result is above IDR 30 million, then the SDGs Desa target has been achieved. The formula is as follows:

$$\textit{Village GDP per capita} = \frac{\text{Village GDP}}{\sum \text{Village residents}}$$

The target average village GDP per capita is aimed to be above IDR 30 million. The village indicators used are the amount of budget revenue from the Village Revenue and Expenditure Budget (APB Desa) and the amount of budget expenditure from APB Desa. The Neighbourhood Units/hamlets (Rukun Tetangga/RT) indicators used are the amount of profit of each business in the village and the amount of expenditure of each business in the village. Individual indicators used were the amount of income of each family member in the previous month and the amount of expenditure of each family member in the same month. Activities that can be done include: mapping village potential, and developing village potential through development of tourist villages, MSME centres, and village products. Tourism development in the village can be done by building tourist attraction facilities, forming Tourism Awareness Groups (Kelompok Sadar Wisata/Pokdarwis), and increasing the capacity of Pokdarwis/other tourism actors. The development of village MSME centres can be done by providing capital for MSME actors, and providing training related to marketing MSME products through e-commerce. The development of flagship village products can be done by providing assistance in processing flagship village commodities, building infrastructure support to increase the productivity of flagship commodities such as irrigation, roads, bridges, facilities, and, post-harvest, utilising vacant land to increase the production of flagship commodities, optimising social forestry programmes, digitising the marketing of flagship village products, and mentoring online store of Village-owned Enterprises (BUM Desa).

13.3.2 Village GDP per Worker in the Village

To measure the village GDP per worker in the village, the village GDP is needed, which is the total of village funds (Dana Desa), PADes, local tax revenues, Village Fund Allocation (ADD), financial assistance from the Provincial APBD, financial assistance from the APBD district/city, grants, and others. This figure is then divided by the number of workers in the village. If the result is above IDR 30 million, then the SDGs Desa target has been achieved. The formula is as follows:

$$\textit{Village GDP per worker} = \frac{\text{Village GDP}}{\sum \text{Workers in the village}}$$

The target average per capita village GDP is aimed to be above IDR 30 million per worker in the village. The village indicators used are the amount of APB Desa income budget and the amount of expenditure budget. The Neighbourhood Units/hamlets (Rukun Tetangga/RT) indicators used are the amount of profit of each business in the village and the amount of expenditure of each business in the village. The individual indicators used are the amount of income of each family member in the most recent month, the amount of expenditure of each family member in the previous month, and the number of family members who work.

13.3.3 Workers in the Formal Sector

To measure the percentage of workers in the formal sector, the number of people working as teachers, industrial workers, civil servants, military personnel, village apparatus, and village office employees aged over 15 years is divided by the number of working people over the age of 15, then multiplied by 100%. If the result is above 51%, then the SDGs Desa target has been achieved. The formula is as follows:

$$\textit{\% Formal workers} = \frac{\sum \text{Population with formal employment}}{\sum \text{Population working aged} > 15 \text{ years}} \times 100\%$$

The target percentage of workers in the formal sector is set at a minimum of 51%. The individual indicator used is employment over the past year and the year before. Activities that can be carried out include: conducting socialisation of the formal sector job market for the village workforce, as well as conducting training and certification of formal sector workforce.

13.3.4 Micro, Small and Medium Enterprises (MSMEs) Have Access to Formal Capital

To measure the percentage of MSMEs that have access to formal capital in each Neighbourhood Units /hamlets (Rukun Tetangga/RT) area, the number of MSME actors who can get capital loans from financial institutions and the ease of borrowing capital at financial institutions (easy, moderate, difficult), is divided by the number of MSME actors, then multiplied by 100%. If the result is above 51%, then the SDGs Desa target has been achieved. The formula is as follows:

$$\%\ \textit{MSMEs that have access to formal capital} = \frac{\sum \text{MSME actors who get capital loans from financial institutions}}{\sum \text{MSME actors}} \times 100\%$$

The target for micro, small and medium enterprises (SMEs) in the village who have access to formal capital is set at a minimum of 51%. The individual indicator used is the number of capital loans at financial institutions and the ease of borrowing capital at financial institutions (easy, moderate, difficult). Activities that can be carried out include: conducting socialisation of financial access expansion and other capital for SME operators, providing capital assistance managed by Village-owned Enterprises (BUM Desa), and accompanying SME business groups in the utilisation of business capital.

13.3.5 Number of People Working in the Workforce Group

To measure the number of people working, including the workforce group, data on the number of people working, including the workforce group is needed. This number is divided by the number of people included in the workforce group, then multiplied by 100%. If the result is 100%, then the SDGs Desa target has been achieved. The formula is as follows:

$$\%\ \textit{People working in the Workforce Group} = \frac{\sum \text{People working in the workforce group}}{\sum \text{People in the workforce group}} \times 100\%$$

The target number of people working is set at 100% of the workforce. The individual indicators used are the jobs of each family member over the past year and the year before. Activities that can be carried out include optimising the Cash-for-Work (Padat Karya Tunai Desa/PKTD) in developing productive economic businesses of the community, as well as conducting training in skills such as sewing, arts and crafts, batik, weaving, and others.

13.3.6 Open Unemployment Rate in the Village

To measure the open unemployment rate, data on the number of people who are looking for work aged over 15 years is needed. This number is divided by the number of people aged over 15 years, then multiplied by 100%. If the result is 0%, then the SDGs Desa target has been achieved. The formula is as follows:

$$\%\ \textit{Open Unemployment Rate} = \frac{\sum \text{People who are looking for work aged} > 15 years}{\sum \text{ Population aged} > 15 years} \times 100\%$$

The target open unemployment rate in the village is set at 0%. The individual indicators used are the jobs of each family member over the past year and the year before. Activities that can be carried out include optimising the Cash-for-Work (Padat Karya Tunai Desa/PKTD) in the development of productive economic community businesses, as well as conducting training such as sewing training, craft arts, batik, weaving, and others.

13.3.7 Absorption of Labour in Cash-for-Works Programme Represents Unemployment in the Village

To measure the percentage of labour absorption in PKTD represents unemployment in the village, the number of unemployed who participate in cash-for-work programme (Padat Karya Tunai Desa/PKTD) is divided by the number of cash-for-work programme (PKTD) workers, then multiplied by 100%. If the result is more than 50%, then the SDGs Desa target has been achieved. The formula is as follows:

$$\%\ \textit{Absorption of Labour in cash} - \textit{for} - \textit{workprogramme (PKTD) Represents Unemployment in the Village} = \frac{\sum \text{Unemployed individuals participating in cash-for-work programme (PKTD)}}{\sum \text{cash-for-work programme (PKTD) Workers}} \times 100\%$$

The target for employment absorption in cash-for-work programme (PKTD) is set to be more than 50% of the unemployed individuals in the village. The Neighbourhood Units/hamlets (Rukun Tetangga/RT) indicators used are the number of cash-for-work programme (PKTD) activities, the number of cash-for-work programme

(PKTD) workers, and the number of unemployed workers. Activities that can be carried out include socialising cash-for-work programme (PKTD) to unemployed residents, as well as increasing wages without neglecting the quality of the facilities built.

13.3.8 New Workforce Members Who Have Not Worked and Receive Skills Training

To measure this, the number of new workforce members who have not worked and received skills training is divided by the number of new workforce members who have not worked, then multiplied by 100%. If the result is 100%, then the SDGs Village target has been achieved. The formula is as follows:

$$\%\ \textit{New workforce members who have not worked}\ \text{and}\ \textit{receive skills training} = \frac{\sum \text{New workforce members who have not worked receive skills training}}{\sum \text{New workforce members who have not worked}} \times 100\%$$

The target for new workforce members who have not worked to receive skills training is set at 100%. The village indicator used is the number of job skills training courses/events for residents. The individual indicator used is the number of job skills training courses/events attended. Activities that can be carried out include conducting potential job market socialisation for the village workforce, conducting campaigns on the importance of vocational/technical education, preparing skilled workers through training and certification of expertise, conducting entrepreneurship training and other themed courses, forming SME groups in the village, and providing business group assistance from formation, organisational training, potential analysis, proposal of productive business activities, implementation of activities, financial and capital access, capital management, and packaging to marketing.

13.3.9 Acts of Violent Abuse, Exploitation, Neglect, and Other Wrongful Actions Towards Children

To measure the management of the mistreatment of children, the number of institutions serving the needs of children in the hamlets (Rukun Tetangga/RT) area is needed. If there are such institutions, then the SDGs Desa target is achieved. The formula is as follows:

$$\textit{Institutions serving the needs of children} = \sum \text{institutions serving the needs of children}$$

In addition, it can be measured by taking the total number of acts of violence, exploitation, neglect, and other wrongful actions towards children and dividing by the total number of acts of violence, exploitation, neglect, and other wrongful actions overall, then multiplying by 100%. If the result is 100%, then the SDGs Desa target is achieved. The formula is as follows:

$$\%\ \textit{Acts of violence, exploitation, neglect, and other wrongful actions towards children} = \frac{\sum \text{Acts of violence, exploitation, neglect, and other wrongful actions towards children}}{\sum \text{Acts of violence, exploitation, neglect, and other wrongful actions}} \times 100\%$$

The target for acts of violence, exploitation, neglect, and other wrongful actions towards children is set at 0%. The Neighbourhood Units/hamlets (Rukun Tetangga/RT) indicator used is the availability of institutions that serve the needs of children. The individual indicators used are age (date of birth) of family members, the number of children who have experienced physical violence, the age of children who are already working, and the number of children who are not cared for. Activities that can be carried out include conducting campaigns and promoting children's rights, child-rearing skills, child protection, and prevention of child marriage, organising community forums for the preparation of proposals for marginal and vulnerable groups, cooperating with social institutions, such as Australia-Indonesia Partnership for Gender Equality and Women's Empowerment (Kemitraan Australia-Indonesia untuk Kesetaraan Gender dan Pemberdayaan Perempuan/MAMPU) (Indonesia-Australia cooperation) and PEKKA, and implementing Community-Based Integrated Child Protection (Perlindungan Anak Terpadu Berbasis Masyarakat/PATBM).

13.3.10 Workplaces with Health and Safety Facilities

To measure the percentage of workplaces that have health and safety facilities, the number of workplaces that have health and safety facilities is divided by the number of workplaces in the village, then multiplied by 100%. If the result is 100%, then the SDGs Desa target has been achieved. The formula is as follows:

$$\%\ \textit{Workplaces with health and safety facilities} = \frac{\text{Workplaces with health and safety facilities}}{\sum \text{Workplaces in the village}} \times 100\%$$

It can also be measured by summing up the village residents who receive health and safety guarantees from their workplace. This total is divided by the number of residents who work, then multiplied by 100%. If the result is 100%, then the SDGs Desa target has been achieved. The formula is as follows:

$$\%\ \textit{Village residents receiving health and safety facilities in the workplace} = \frac{\sum \text{Village residents receiving health and safety guarantees from the workplace}}{\sum \text{Working village residents}} \times 100\%$$

The target for workplaces to have health and safety facilities is set at 100%. The Neighbourhood Units/hamlets (Rukun Tetangga/RT) indicator used is the availability of health facilities, health insurance, and safety facilities in the workplace. The individual indicator used is the number of family members who receive health and safety guarantees from the workplace. Activities that can be carried out include implementing a Covid-19 Safe Village in the village environment, including in potential areas for the development of productive economic activities of citizens (for example, tourist areas).

13.3.11 Contribution of Tourism Activities to Village GDP

To measure the economic contribution of tourism, the total village GDP that comes from the turnover of village-owned tourist spots for one year is divided by the village GDP for one year, then multiplied by 100%. If the result is more than 8%, then the SDGs Desa target has been achieved. The formula is as follows:

$$\%\ \textit{Contribution of tourism activities to village GDP} = \frac{\text{Village GDP that comes from the turnover of village} - \text{owned tourist spots}}{\text{Village GDP}} \times 100\%$$

The target for contribution of tourism activities is set at a minimum of 8% of the village GDP. The Neighbourhood Units/hamlets (Rukun Tetangga/RT) indicator used is the number of tourist spots owned by the village, last year's tourism turnover, this year's tourism turnover, last year's tourism profit, and this year's tourism profit. Activities that can be carried out include: construction, maintenance, and management of tourist attraction facilities in the village; improving the capacity of tourism actors in the village (through training and certification of hospitality, culinary, herbal processing, and weaving); digitising tourism (in transportation accommodation, lodging, objects, and others); working with third parties for village tourism investment, and promoting village tourism through cultural exhibitions and digital-based booking and information systems.

13.3.12 Number of Tourists

To measure the percentage increase in the number of tourists, the number of tourists in a certain year, minus the number of tourists the previous year, is divided by the number of tourists the previous year, then multiplied by 100%. If the result is more than 0% (indicating an increase), then the SDGs Desa target has been achieved. The formula is as follows:

$$\% \, Increase \, in \, the \, number \, of \, tourists = \frac{\sum \text{Tourists in a certain year} - \sum \text{Tourists in the previous year}}{\sum \text{Tourists in the previous year}} \times 100\%$$

The target number of tourists is aimed to increase each year. The Neighbourhood Units/hamlets (Rukun Tetangga/RT) indicators used is the number of domestic and international tourists in a certain year and the previous year. Activities that can be carried out include: building tourist attractions to increase tourist appeal, as well as digitising tourism to make it easier for tourists to access tourist information and accommodation.

13.3.13 Access to Capital and Financial Services

To measure access to capital and financial services in each Neighbourhood Units/hamlets (Rukun Tetangga/RT) area, the number of institutions that can assist with SME capital, such as the number of Village Cooperative Units (Koperasi Unit Desa/KUD) providing business credit and the number of savings and loan cooperatives in the Neighbourhood Units/hamlets (Rukun Tetangga/RT) area, is divided by the number of Neighbourhood Units/hamlets (Rukun Tetangga/RT), then multiplied by 100%. If the result is above 0%, then the SDGs Desa target has been achieved. The formula is as follows:

$$\% \, Access \, to \, capital \, and \, financial \, services \, in \, each \, hamlet (RT) \, area$$
$$= \frac{\sum \text{Village Cooperative Units and Savings and Loan Cooperatives in each hamlet (RT)}}{\sum \text{hamlets (RT)}} \times 100\%$$

This target aims to provide access to capital and financial services. The indicator used for the Neighbourhood Units/hamlet (Rukun Tetangga/RT) is the number of financial institutions in the Neighbourhood Units/hamlets (Rukun Tetangga/RT) area. Activities that can be carried out include socialising the expansion of financial access and other capital for MSME actors, providing capital assistance managed by the Village-owned Enterprises (BUM Desa), and providing assistance to MSME business groups in the utilisation of business capital.

13.4 Operational Definitions for Equitable Village Economic Growth

Key words related to Equitable Village Economic Growth are presented in Table 13.1.

Table 13.1 Operational definitions for equitable village economic growth

	Key word	Operational definition
1.	Workforce	Working-age population (15 years or older) who are working, or have jobs but are temporarily not working, or are unemployed
2.	Village-owned enterprises (BUM Des)	Village businesses with legal entity status managed by the village government according to the needs and potential of the village
3.	Village funds	Funds allocated from the state budget intended for villages that are transferred through the Regional Revenue and Expenditure Budget and used to finance the administration of government, implementation of development, community development, and community empowerment
4.	COVID-19 safe village	Efforts to handle COVID-19 by forming village volunteers against COVID-19 using the principle of mutual cooperation involving the support of village community members
5.	*E-commerce*	Website used to sell products online
6.	Tourist village	A community consisting of residents of a limited area acting directly in efforts to increase readiness and concern in responding to tourism potential or tourist attraction locations in each village area
7.	Village Unit Cooperative (KUD)	Cooperatives in rural areas that engage in providing community needs
8.	Open unemployment	People who do not have a job and are looking for work; do not have a job and are preparing a business, do not have a job and are not looking for work, because they feel it is impossible to get a job; or already have a job, but haven't started working yet
9.	Monthly income (IDR)	Receipt of money for activities carried out or services provided in one month in rupiah units
10.	Monthly expenditure (IDR)	Disbursement of money for goods or services obtained in one month in rupiah units
11.	Economic infrastructure	Tools to directly or indirectly support all types of facilities, such as public facilities, roads, bridges, etc.
12.	Village fund programme	Programmes/activities carried out in the village/subdistrict from the allocation of village funds
13.	Program activities from the allocation of village funds	Programme activities in the village/subdistrict regulated in a government policy (ministerial regulation, decree, etc.)
14.	Village production potential	The ability, capability that can be produced from the village
15.	Tourism potential	Everything owned for tourist attraction
16.	Gross domestic product (GDP) of the village	The value of goods and services produced by the village
17.	Economic facilities	Tools that can be used to facilitate or ease economic activities, such as markets, cars, public vehicles, etc.

(continued)

Table 13.1 (continued)

	Key word	Operational definition
18.	Online shop online	Media for buying and selling products through electronic media
19.	Micro, small, medium enterprises (MSME)	Businesses run by individuals, households, or small-sized business entities

Notes

Vocab: PKTD: Cash-for-Work, reference: https://www.tnp2k.go.id/downloads/prioritising-village-fund-utilisation:-cash-for-work-to-prevent-covid-19.

Reference

World Bank. 2005. *World development report 2006 : Equity and development*. Washington DC (USA): World Bank.

Chapter 14
SDGs Desa 9: Needs-Based Village Infrastructure and Innovation

14.1 Concept of Needs-Based Village Infrastructure and Innovation

Village infrastructure is the physical facilities of the village, including the existence and condition of roads, bridges, irrigation, and public facilities to assist community access to socio-economic systems. Further development of the infrastructure is important in promoting the growth of the village area (Kusuma 2011). The development of village infrastructure can enhance the welfare of the village community and the quality of human life, as well as alleviate poverty through the fulfilment of basic needs, the development of village facilities and infrastructure, the development of local economic potential, and the sustainable use of natural resources and the environment. Therefore, development of the infrastructure is essential. To achieve positive economic growth, the reliability of village infrastructure is crucial. This includes village road infrastructure, such as main roads, as well as other infrastructure that supports the economic activities of village residents, such as in the agricultural, fisheries, and other sectors.

SDGs Desa 9 also emphasises the beginning of village innovation. Village innovation is any activities that develop the potential in the village in a new way or produce new products. It focuses especially on developing community capacity, for example creating new entrepreneurs, producing outstanding village products, improving the quality of public services, and increasing the economy through the village GDP. Therefore, SDGs Desa use several success indicators that can be carried out by the village government and supra village, including reliable road conditions, docks/boat moorings, the growth of industry in the village, and the contribution of industry to village economic growth.

A. H. Iskandar, *The Methods of Measuring the National Sustainable Development Goals in Indonesia*, Engaging Indonesia, https://doi.org/10.1007/978-981-96-0927-7_14

14.2 Needs-Based Village Innovation and Infrastructure Targets

To achieve the goal of Needs-based Village Innovation and Infrastructure, the following six targets have been set to be achieved by the year 2030:

Good road conditions reach 100%

(1) Good dock/boat mooring conditions reach 100%
(2) The growth rate of household, small and medium industries is faster than village GDP
(3) The contribution of processing industry activities reaches at least 8% of village GDP
(4) The percentage of industries that pollute the air reaches 0%
(5) Village innovations that are recorded and disseminated reach 100%.

14.3 Measurement and Analysis of Needs-Based Village Innovation and Infrastructure

14.3.1 Good Road Conditions

To measure the percentage of good road conditions, the number of Neighbourhood Units/hamlets (Rukun Tetangga/RT) with roads that can be used by 4-wheel drive vehicles is divided by the number of hamlets (Rukun Tetangga/RT), then multiplied by 100%. If the result is 100%, then the SDGs Desa target is achieved. The formula is as follows:

$$\%\ \text{Good Road Conditions} = \frac{\sum \text{Length of roads (km) in good condition}}{\sum \text{Length of roads (km)}} \times 100\%$$

The target for good road conditions is set at 100%. Neighbourhood Units Indicator (hamlet/Rukun Tetangga/RT) used are road lengths by type (asphalt/gravel or stone/soil/wooden over water/footpath/others) and road conditions (good, moderate, poor). Activities that can be carried out include mapping the condition and development needs of roads, as well as carrying out the construction and improvement of neighbourhood roads, local roads, strategic village roads, and agricultural roads.

14.3.2 Good Condition of Dock/Boat Moorings

To measure the percentage of docks/boat moorings in good condition, the number of sea/river docks in good condition is divided by the total number of sea/river docks, then multiplied by 100%. If the result is 100%, then the SDGs Desa target is achieved. The formula is as follows:

$$\% \text{ Good Condition Dock/Boat Mooring} = \frac{\sum \text{Sea/river docks in good condition}}{\sum \text{Sea/river docks}} \times 100\%$$

The target for docks/boat moorings in good condition is set at 100%. The Neighbourhood Units/hamlets (Rukun Tetangga/RTs) indicator used is the number of boat moorings and the condition of the boat moorings (good, average, poor). Activities that can be carried out are the construction or rehabilitation of public docks/boat moorings.

14.3.3 Growth Rate of Household, Small, and Medium Industries Compared with Village GDP

To measure the growth rate of household, small, and medium industries, data on the total value of final goods and services produced by household, small, and medium industries in a certain year is needed, minus the total value of final goods and services produced by household, small, and medium industries in the previous year. This number is divided by the total value of final goods and services produced by household, small, and medium industries in the previous year, then multiplied by 100%. If the result is higher than the growth rate of the village GDP, then the SDGs Desa target is achieved. The formula is as follows:

$$\text{Growth Rate of Household, Small, and Medium Industries} = \frac{\text{Total value of final goods and services in a certain year} - \text{total value of final goods and services in the previous year}}{\text{Total value of final goods and services in the previous year}} \times 100\%$$

The growth rate target for household, small, and medium industries is set to be faster than the village GDP. The Neighbourhood Units/hamlets (Rukun Tetangga/RTs) indicator used is the type of household, small, and medium industries, the turnover of household, small, and medium industries, and the profits of household, small, and medium industries. Activities that can be carried out include: conducting a census of micro- and medium-scale processing industry players in the village, forming a processing industry cluster in the village, providing capital assistance for village micro- and medium-scale industry groups, conducting training and mentoring for packaging of products, and carrying out digitalisation of product-marketing MSMEs (online shop, e-commerce) managed by Village-owned Enterprises.

14.3.4 Contribution of Processing Industry Activities to Village GDP

To measure the percentage contribution of processing-industry activities to the village GDP, data is needed on the village GDP that comes from processing industries (such as leather, wood, precious metals, heavy metals, cloth, pottery/ceramics/stone, tiles/ bricks, weaving, food/beverages). This figure is divided by the total village GDP, then multiplied by 100%. If the result is more than 8%, then the SDGs Desa target has been achieved. The formula is as follows:

$$\begin{aligned}&\%\text{ Contribution of Processing Industry Activities to Village GDP}\\&=\frac{\text{Village GDP derived from the processing industry}}{\text{PDB Desa}}\times 100\%\end{aligned}$$

The target contribution of processing industry activities is set at a minimum of 8% of the village GDP. The Neighbourhood Units/hamlets (Rukun Tetangga/RTs) indicator used is the types of household, small, and medium industries, the turnover of household, small, and medium industries and the profits of household, small, and medium industries.

14.3.5 Percentage of Industries that Pollute the Air

To measure the percentage of industries that pollute the air, the number of industries that pollute the air is divided by the number of industries, then multiplied by 100%. If the result is 0%, then the SDGs Desa target is achieved. The formula is as follows:

$$\%\text{ Air-Polluting Industries}=\frac{\sum\text{Industries that pollute the air}}{\sum\text{Industries}}\times 100\%$$

The target percentage of industries that pollute the air is set at 0%. The Neighbourhood Units/hamlets (Rukun Tetangga/RTs) indicator used is a list of companies, and a list of companies that pollute the air. Activities that can be carried out include: conducting campaigns for the use of environmentally friendly technology, initiating movements for the use of environmentally friendly technology, utilising social forestry for nature tourism in collaboration with Village-owned Enterprises, carrying out reforestation, as well as building active and passive parks in the village.

14.3.6 Village Innovations Recorded and Disseminated

To measure the percentage of village innovations that are recorded and disseminated to social media, data is needed on the existence of village websites, village Facebook pages, village Twitter accounts, village Instagram accounts, and village YouTube channels. The existence of each type of village social media is given one point. These points are added up, divided by five, then multiplied by 100%. If the result is 100%, then the SDGs Desa target is achieved. The formula is as follows:

$$\begin{aligned}&\%\ \text{Existence of Village Social Media}\\&=\frac{\text{Availability of village website Facebook, Twitter, Instagram, Youtube}}{5}\times 100\%\end{aligned}$$

This target aims for the existence of village innovations that are recorded and disseminated. The village indicators used are the village website, village Facebook pages, village Twitter accounts, village Instagram accounts, and village YouTube channels. Activities that can be carried out include disseminating village innovations through social media regularly, as well as improving human resource capacity to disseminate village innovations in an interesting way.

14.4 Operational Definitions for Needs-Based Village Innovation and Infrastructure

Key words related to Needs-based Village Infrastructure and Innovation are presented in Table 14.1.

Table 14.1 Operational definitions for needs-based village innovation and infrastructure

	Keyword	Operational definition
1	Dock/boat mooring	A place to dock and moor boats that load and unload goods, embark and disembark people, and so on
2	Facebook	A social network that allows users to add other users as friends, exchange messages, join user groups with similar interests, and so on
3	Instagram	A social network that allows users to share photos and short duration videos online
4	Road	Land transportation infrastructure that includes all parts of the road, including complementary buildings and their equipment intended for traffic, which is on the surface of the earth, above the surface of the earth, above the surface of the water and under the surface of the earth and or water, except for railway tracks, lorry roads and cable roads
5	Village road	A public road that connects areas and/or settlements within the village, as well as neighbourhood roads within the village
6	Village axis road	A road that connects two villages
7	Village strategic road	A road that is prioritised to serve the interests of several villages based on considerations to stimulate economic growth, welfare, village security
8	Twitter	A social network that allows users to read and create short messages called tweets
9	YouTube	A social network that allows users to share long-duration videos online
10	Website	A collection of pages on the internet to display information online

Reference

Kusuma, W. 2011. Village government strategy in efforts to increase village infrastructure development [thesis]. Faculty of Social Sciences and Political Sciences, Muhammadiyah University of Malang.

Chapter 15
SDGs Desa 10: Village Without Gaps

15.1 Concept of Village Without Gaps

SDGs Desa 10 aims to reduce and eliminate disparities in villages by 2030. Eliminating disparities means ensuring no village community is left behind. In developing countries, disparities have increased up to 11%. Currently, 1% of the world's richest population controls up to 40% of global assets, while 10% of the world's poor population only has 2–7% (Oxfam 2016). Disparity is the inequality of access to economic resources (Syawie 2011). Based on the data of the percentage of poor population by region, the percentage of poor population in rural areas is higher than in urban areas (Fig. 15.1).

The widening gap is a sign of the need to empower citizens with the lowest incomes and promote economic inclusion regardless of gender, race or ethnicity. The success of achieving this goal can be measured by the village's Gini coefficient, the level of poverty in the village, the status of village development, and the civil liberty index in the village.

15.2 Village Without Gaps Targets

To achieve the Village without Gaps goal, the following seven targets have been set to be achieved by 2030:

(1) Gini coefficient below 0.200
(2) Poverty rate 0%
(3) Independent Village Status
(4) Civil Liberty Index reaches 100

A. H. Iskandar, *The Methods of Measuring the National Sustainable Development Goals in Indonesia*, Engaging Indonesia, https://doi.org/10.1007/978-981-96-0927-7_15

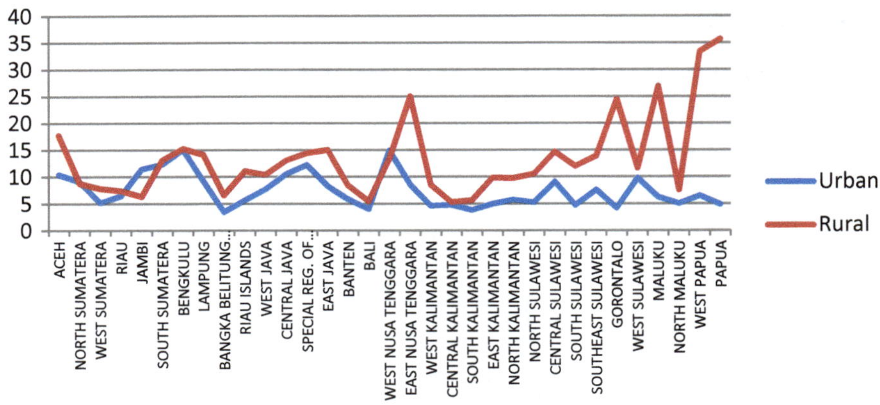

Fig. 15.1 Percentage of poor population by province and region as of March 2021. *Source* BPS (2021)

(5) Number of workers participating in the National Social Security System in employment reaches 100%
(6) Violence against migrant workers reaches 0%
(7) Availability of advocacy rules for prospective and migrant workers.

15.3 Measurement and Analysis of Village Without Gaps

15.3.1 Gini Coefficient

To calculate the Gini coefficient, expenditure data divided into several classes is needed. The formula is as follows:

$$Gini\ coefficient = 1 - \sum fpi \times (F_{ci} + F_{ci-1})$$

Explanation:

fpi = frequency of population in expenditure class i

F_{ci} = cumulative frequency of total expenditure in expenditure class i

F_{ci-1} = cumulative frequency of total expenditure in expenditure class $(i-1)$

If the result is below 0.200, the SDGs Desa target is achieved. The Gini coefficient target is set below 0.200. The individual indicators used are the number of family members and the total income of family members. Efforts that can be made to increase the income of the lowest population include: involving the lowest community in village infrastructure development work and providing daily wages such as through Cash-for-Work (PKTD), opening up job opportunities by creating village business units, providing human resource skills improvement training, such as workshop training or processing of superior commodities, and providing unconditional cash transfers.

15.3.2 Poverty Level

Before measuring the poverty level, first measure the per capita income per day. Per capita income per day is calculated from the income of all individuals in one year divided by the number of individuals, then divided again by 365. The formula is as follows:

$$Income\,per\,capita\,per\,day\;(Rp) = \frac{\sum_{i=1}^{Number\;of\;Individuals} \text{Income per year}}{\sum \text{Individuals}} \times \frac{1}{365}$$

The World Bank sets the standard for extreme poverty line at 1.9 dollars PPP (Purchasing Power Parity), which indicates the number of rupiah spent to buy a certain amount of goods and services based on the price and quantity in each country. This is equivalent to IDR 11,000 per capita per day. This means, if the per capita income per day is below IDR 11,000, then that person is considered extremely poor. Values below IDR 11,000 are included in the calculation of the extreme poverty rate.

The extreme poverty rate or the number of extremely poor citizens is obtained from the number of citizens with per capita income per day that is below IDR 12,000, divided by the number of individuals, multiplied by 100%. If the result is 0%, it can be said that the SDGs Desa target has been achieved. The formula used is as follows:

$$Poverty\;Rate = \frac{\sum_{i=1}^{Number\;of\;Individuals} Individuals\;with\;income\;per\;capita\;per\;day \le IDR 12.000}{\sum \text{Individuals}} \times 100\%$$

The poverty rate target is set at 0%. The individual head indicators used are home ownership status (owned/rented), home structure (bamboo, wood, brick, soil, tile, cement, ceramic), as well as food and non-food expenditure in the last month. Activities that can be carried out in the village to achieve a village poverty rate of 0% include increasing productivity and added value of commodities through reducing production input costs and financing spent to gain access to public services. For example, building a village market, village online shop, storage warehouse, and other things related to marketing-support facilities. Another activity is developing village food reserves, whether it's rice, corn or sorghum through the village granary.

15.3.3 Independent Village Status

The village status target is set to be 'independent'. The village indicators that the tool used is the Village Development Index (Indeks Desa embangun/IDM). The IDM consists of three composite indices, the Social Resilience Index, the Economic Resilience Index, and the Village Ecological Resilience Index. Achieving a self-reliant village can be done by improving the quality of the village from these three aspects. For the aspect of social resilience, this means increasing social solidarity, tolerance, a sense of security, social welfare, health services, community empowerment, health insurance, and access to education, clean water, sanitation, electricity, and information and communication.

The way to improve economic resilience is through diversification of village community production, providing trade service centres, expanding distribution channels, facilitating access to financial and credit institutions, ensuring the role of economic institutions, and fostering regional openness. For the aspect of environmental resilience, this means improving environmental quality, identifying potential disaster-prone areas and conducting disaster response activities.

15.3.4 Civil Liberty Index

To measure the Civil Liberty Index, the following formula can be used:

$$\textit{Civil Liberty Index} = P \text{ x } I(A)$$

Explanation:

P = Weight Value based on Analytical Hierarchy Process (AHP) from the aspect of civil liberty

$I(A)$ = Index of civil liberty aspect

The result of the index is a scale of 1–100, that is, <60: poor, 60–80: average, and >80: good. If the result is 100, it can be said that the SDGs Desa target has been achieved. The Civil Liberty Index target is set at 100. The individual indicator used is the level of village openness to citizen input over the past year (good, sufficient, poor). Efforts that can be made so that the community is free to express opinions or provide input in the village activity plan include creating a suggestion box/input for the general public, holding village deliberation forums open to all residents, and creating integrated village complaint service posts.

15.3.5 Number of Workers Participating in the National Social Security System for Employment Sector

To measure the percentage of workers participating in the National Employment Social Security (BPJS Ketenagakerjaan), the number of National Employment Social Security participants aged over 15 years is divided by the number of workers aged over 15 years, then multiplied by 100%. If the result is 100%, then the SDGs Desa target is achieved. The formula is as follows:

$$\%National\ Employment\ Social\ Security = \frac{\sum \text{Participants in the national employment social security program aged} > 15\text{ years}}{\sum \text{Workers aged} > 15\text{years}} \times 100\%$$

The target number of workers participating in the National Social Security System for employment is set at 100%. The individual indicator used is the number of family members who are participants in the social security for employment. Efforts that can be made to encourage all working residents to have National Employment Social Security Cards include: conducting a census of residents who do not yet have an employment guarantee card, socialising the use and ownership of the employment guarantee card, and proposing the employment guarantee card for villagers who need it.

15.3.6 Violence Against Migrant Workers

To measure the percentage of violence against migrants, the number of violent acts against migrants is divided by the total number of violent acts in the village, then multiplied by 100%. If the result is 0%, then the SDGs Desa target is achieved. The formula is as follows:

$$\%Violence\ against\ migrant\ workers = \frac{\sum \text{Violence against migrant workers}}{\sum \text{Violence in the village}} \times 100\%$$

The target for violence against migrant workers is set at 0%. The village indicators used are the existence of village cooperation with third parties regarding the protection of migrant workers. The individual indicators used are the number of family members who are migrant workers, the number of these migrant workers who experience violence at work, and the number of migrant workers who receive protection services. Efforts that can be made to eliminate violence against migrant workers include: implementing cooperation agreements with official agents and formal parties related to the dispatch of migrant workers abroad so that there is a Memorandum of Understanding (MoU) with the relevant parties regarding the law and protection of

Table 15.1 Operational definitions for a village without gaps

	Key word	Operational definition
1.	Civil freedom index	A value indicating freedom of religion and belief, expression, association, and opinion
2.	Gini coefficient	A method of calculation related to the income distribution of a population.
3.	Migrant workers	People who work abroad

migrant workers. Other actions include: establishing NGOs or shelters for the children of migrant workers, establishing integrated complaint services that can provide advocacy and mediation for the problems of migrant workers in the village, and carrying out data collection of prospective migrant workers, including the legality of documents and agents and checking departure procedures.

15.3.7 Availability of Advocacy Rules for Prospective and Migrant Workers

To measure the availability of advocacy rules for prospective and migrant workers, a village regulation/village head decree is needed regarding advocacy for prospective and migrant workers. If there are rules related to this, then the SDGs Desa target is achieved.

$$\textit{Advocacy Rules for Prospective and Migrant Workers} = \sum \begin{matrix} \text{Village regulations or village head regulations or decrees} \\ \text{regarding advocacy for prospective and migrant workers} \end{matrix}$$

This target aims for the existence of advocacy rules for candidates and migrant workers. The village indicator used is the existence of Village/Village/Head of Village regarding advocacy for candidate and migrant workers. The individual indicator used is the number of family members who become migrant workers and the number of family members of migrant workers who are placed.

15.4 Operational Definitions for Village Without Gaps

Key words related to the Village without Gaps are presented in Table 15.1.

References

BPS. 2021. Poverty profile in Indonesia. In *Official statistics news*, vol. 24, no. 16/02.
Oxfam. 2016. *An economy for the 1%* [internet]. Available at: https://oxfamilibrary.openrepository.com/bitstream/handle/10546/592643/bp210-economy-one-percent-tax-havens-180116-en.pdf;jsessionid=C2A90B727EA39902E8F8F0464D91F304?sequence=47.
Syawie M. 2011. Poverty and social gap. In *Information*, vol. 16, no. 03.

Chapter 16
SDGs Desa 11: Safe and Comfortable Village Residential Areas

16.1 Concept of Safe and Comfortable Village Residential Areas

A settlement area is part of the environment outside the protected area, whether urban or rural, which functions as a residential environment or living environment and a place of activity to support life and livelihood and, essentially, to realise liveable, safe, comfortable, peaceful, prosperous, and sustainable urban and rural settlements (KemenPUPR 2020).

This goal, by 2030, targets the realisation of an inclusive, safe, resilient, and sustainable village, with several achievement targets for clean and healthy settlement areas, the creation of environmental security through community self-reliance, and the participation of all parties in village development. Creating a safe and comfortable village settlement area means ensuring access to safe and affordable housing, as well as improving slum settlements. This also involves investing in public transportation, creating public green spaces, and enhancing inclusive village planning and regulation while involving all parties.

16.2 Targets for Safe and Comfortable Village Residential Areas

To achieve the goal of a Safe and Comfortable Village Residential Areas, the following 13 targets have been set to be achieved by 2030:

(1) Slum house percentage reaches 0%
(2) The implementation of security reaches 100% in all Neighbourhood Units/hamlets (Rukun Tetangga/RT)
(3) Families using public transportation exceed 50%

A. H. Iskandar, *The Methods of Measuring the National Sustainable Development Goals in Indonesia*, Engaging Indonesia, https://doi.org/10.1007/978-981-96-0927-7_16

(4) Children, women, people with disabilities, elderly using public transportation exceed 50%
(5) Population moving to the city is less than 15%
(6) There is involvement of private sector, community organisations and professional organisations in social gatherings
(7) Preserved culture reaches 100%
(8) Disaster Risk Index (DRI) reaches 0% in all Neighbourhood Units/hamlets (Rukun Tetangga/RT)
(9) Families receive warnings in disaster-prone areas
(10) Availability of early warning systems for weather and climate as well as disasters
(11) Household waste management reaches 100%
(12) There is a waste-processing facility in the village
(13) Presence of parks/fields in the village.

16.3 Measurement and Analysis of Safe and Comfortable Village Residential Areas

16.3.1 Percentage of Slum Houses

Slum households are defined as households that do not have access to decent drinking water sources, do not have access to decent sanitation, do not have access to floor area >= 7.2 m^2 per capita, and do not have access to decent roof, floor, and wall conditions (BPS 2019). To measure the percentage of slum houses, the number of slum houses is divided by the number of houses in the village, then multiplied by 100%. If the result is 0%, then the SDGs Desa target is achieved. The formula is as follows:

$$\%\,Slum\,House\,Conditions = \frac{\sum \text{Slum Houses}}{\sum \text{Family Houses}} \times 100\%$$

The target percentage of slum houses is set at 0%. The individual indicators used are the status of home ownership (owned, rented/leased, rent-free, borrowed, service, or other), the status of the land where the house is located (owned, owned by others, state land, other), the structure of the house (floor, walls, windows, roof), and the condition of the house's environment. Efforts that can be made so that all villagers have decent homes include: conducting a census of low-income people whose homes are not decent, building and/or repairing homes for the poor, providing assistance with building materials for light, medium, and heavy rehabilitation of low-income people, renovating the homes of low-income people that are not decent using the Cash-for-Work (PKTD) allowances, and conducting socialisation of the Ministry of Public Works and Public Housing's Decent Home Programme assistance for low-income people.

16.3.2 Security in Neighbourhood Units

To measure the percentage of security activities in Neighbourhood Units/hamlet (Rukun Tetangga/RTs), the number of hamlets (Rukun Tetangga/RT) that have held security activities in the past year is divided by the number of hamlets (Rukun Tetangga/RT), then multiplied by 100%. If the result is 100%, then the SDGs Desa target is achieved. The formula is as follows:

$$\% \, Security \, Activities \, in \, hamlets(RukunTetangga/RT) = \frac{\sum \text{hamlets (Rukun Tetangga/RT) that conducted security activities}}{\sum \text{hamlets (Rukun Tetangga/RT)}} \times 100\%$$

The target for the implementation of security in the hamlets (Rukun Tetangga/RT) environment is set at 100%. The indicators used for the Neighbourhood Units/hamlets (Rukun Tetangga/RT) are the number of security officers, the number of guard posts, the number of people who do night patrols, the existence of mandatory reporting rules, and the socialisation of newcomers. Efforts to improve the system and access to safe settlement services in the village include: establishing a village guard post, building a 24-hour guest reporting room, and regularly implementing the Neighbourhood Watch Programme (Sistem Keamanan Lingkungan/Siskamling): a programme that allows groups of people to police their own neighbourhoods to prevent crime.

16.3.3 Families Using Public Transport

To measure the percentage of families using public transportation, the number of families using public transportation is divided by the total number of families in the village and then multiplied by 100%. If the result is above 50%, then the SDGs Desa target is achieved. The formula is as follows:

$$\% \, \text{Families using public transport} = \frac{\sum \text{Families using public transport}}{\sum \text{Families}} \times 100\%$$

The target for families using public transport is set to be more than 50%. The family indicator used is the number of family members who used public transport in the last month and the month before. Efforts that can be made to provide public transport for villagers to economic centres in the village include buying public transport (like pick-up cars) for the village/county market and providing transport subsidies for people who use public transport.

16.3.4 Children, Women, People with Disabilities, and the Elderly Using Public Transport

To measure the percentage of children, women, people with disabilities, and the elderly who use public transport, the number of children, women, people with disabilities, and the elderly who use public transport is divided by the number of children, women, people with disabilities, and the elderly in the village, then multiplied by 100%. If the result is above 50%, then the SDGs Desa target is achieved. The formulas are as follows:

$$\%\,Children\ using\ Public\ Transport = \frac{\sum \text{Children using Public Transport}}{\sum \text{Children}} \times 100\%$$

$$\%\,Women\ using\ Public\ Transport = \frac{\sum \text{Women using Public Transport}}{\sum \text{Women}} \times 100\%$$

$$\%\,Disabled\ People\ using\ Public\ Transport = \frac{\sum \text{Disabled people using public transport}}{\sum \text{Disabled people}} \times 100\%$$

$$\%\,Elderly\ using\ Public\ Transport = \frac{\sum \text{Elderly using public transport}}{\sum \text{Elderly}}$$

The target for children, women, people with disabilities, elderly public transport users is set at more than 50%. The individual indicators used are the number of family members (children, women, people with disabilities, elderly) who have used public transport in the last month and the month before. Efforts that can be made to provide public transport for children, women, people with disabilities, and the elderly in the village to the village's economic centres include buying public transport (like pick-up cars) to the village/county market and providing transport subsidies for people who use public transport.

16.3.5 Population Moving to the City

To measure the percentage of the population moving to the city, the number of people moving to the city is divided by the number of people in the village, then multiplied by 100%. If the result is below 15%, then the SDGs Desa target is achieved. The formula is as follows:

$$\%\,Population\ moving\ to\ the\ City = \frac{\sum Population\ moving\ to\ the\ city}{\sum Population} \times 100\%$$

The target for the population moving to the city is set to be less than 15%. The Neighbourhood Units/hamlets (Rukun Tetangga/RT) indicator used are the number of people moving to the city and the number of people moving into the village.

The individual indicator used is the family members' place of residence a year ago. Efforts to minimise the migration of village residents to the city include: opening up job opportunities by creating village business units and providing human resource skills enhancement training, such as workshop training or processing of leading commodities. Other efforts include: building village markets, village online stores, storage warehouses, and other marketing support facilities, as well as providing agricultural tools, tractors, vanilla-processing equipment, sewing machines, workshop tools, carpentry tools, or other supporting facilities.

16.3.6 Involvement of Private Roles, Community Organisations and Professional Organisations in Cangkrukan/Jagongan/Kongkow

To measure the involvement of private roles, community organisations and professional organisations in *cangkrukan/jagongan/kongkow* (the activity of gathering together to chat, discuss, and socialise within a village), the number of village collaborations with external parties is needed. If there is involvement from external parties, then the SDGs Desa target is achieved.

The target is for private roles, community organisations and professional organisations to have involvement in *cangkrukan/jagongan/kongkow*. The Neighbourhood Units/hamlets (Rukun Tetangga/RTs) indicator used is the involvement of private roles in *cangkrukan/jagongan/kongkow* and the involvement of village community institutions in *cangkrukan/jagongan/kongkow*.

16.3.7 Preserved Culture

To measure the percentage of preserved culture, the number of existing cultural heritage sites is divided by the total number of cultural heritage sites (both existing and no longer existing), then multiplied by 100%. If the result is 100%, then the SDGs Desa target is achieved. The formula is as follows:

$$\%\,preserved\ cultural\ heritage\ site = \frac{\sum \text{Existing cultural heritage site}}{\sum \text{Cultural heritage sites in the village}} \times 100\%$$

The target for preserved cultural heritage is set at 100%. The Neighbourhood Units/hamlets (Rukun Tetangga/RTs) indicators used are the number of artefacts in the village that are maintained and the number of active village cultures in the past year. Efforts made to preserve historical relics include: building guards for important historical relics, making cooperation agreements related to preservation actions with museums or historical object preservation parties, making customary laws and strict preservation of historical objects, and conducting campaigns and publications related to the guarding of historical relics in the village.

16.3.8 Disaster Risk Index (IRB)

The measurement of the Disaster Risk Index refers to the measurement conducted by the National Disaster Management Agency (BNPB 2020). The data used are (1) hazard per type of disaster, (2) lives exposed per type of disaster, (3) rupiah losses per type of disaster, (4) environmental damage (ha) per type of disaster, and (5) local government capacity per district/city. The determination of the hazard weight per type of disaster is based on the relationship between the frequency of occurrence and the absence of a warning. In the disaster risk level classification Table 16.1, the hazard weight per type of disaster and the hazard level value—1 for low, 2 for medium, and 3 for high—can be seen.

Table 16.2 determines the score based on the exposed lives per type of disaster with an index value of Class 1 for low, Class 2 for medium, and Class 3 for high.

Table 16.3 determines the score based on the loss in rupiah per type of disaster with an index value of: Class 1 for low, Class 2 for medium, and Class 3 for high.

Table 16.4 determines the score based on environmental damage (ha) per type of disaster with an index value of: Class 1 for low, Class 2 for medium, and Class 3 for high.

Table 16.5 determines the score based on the village's capacity with an index value of: 1 for low, 2 for medium, and 3 for high.

Villages can measure the Disaster Risk Index based on the five tables above. The Disaster Risk Index is measured per disaster using the following formula:

$$\textit{Disaster Risk Index X} = \text{Danger score}\ \frac{\textit{Exposed population score} + \textit{Rupiah loss score} + \textit{Environmental damage score}}{\text{Capacity score}}$$

The results of each disaster show the level of disaster risk based on Table 16.6. Then, the overall Disaster Risk Index can be obtained by adding up all the Disaster Risk Indexes from each disaster that has been calculated previously.

The target for the Disaster Risk Index (DRI) is set at 0% in all Neighbourhoods Units/hamlets (Rukun Tetangga/RTs). The Neighbourhood Units/hamlets (Rukun Tetangga/RTs) indicators used are disasters from the past year (floods, landslides, earthquakes, strong winds, tsunamis, fires). The family/individual indicators used are disaster experiences from the past year (floods, landslides, earthquakes, strong winds, tsunamis, fires). Efforts made to build disaster mitigation facilities to minimise impacts include: building evacuation routes, creating potential disaster-prone maps in villages, and providing disaster information services.

Table 16.1 Classification of disaster risk levels (Hazard per Type of Disaster)

No.	Parameter	Value (%)	Class	Weight	Score (Class*Weight)
1.	Earthquake	100	1	4	4
			2		8
			3		12
2.	Tsunami	100	1	4	4
			2		8
			3		12
3.	Volcanic eruption	100	1	3	3
			2		6
			3		9
4.	Flood	100	1	4	4
			2		8
			3		12
5.	Landslide	100	1	5	5
			2		10
			3		15
6.	Drought	100	1	4	4
			2		8
			3		12
7.	Land and forest fires	100	1	4	4
			2		8
			3		12
8.	Extreme weather	100	1	4	4
			2		8
			3		12
9.	Tidal waves/abrasion	100	1	4	4
			2		8
			3		12

16.3.9 Families Receiving Warnings in Disaster-Prone Areas

To measure the percentage of families receiving warnings in disaster-prone areas, the number of families receiving warnings in disaster-prone areas is divided by the number of families in disaster-prone areas, then multiplied by 100%. If the result is 100%, then the SDGs Desa target is achieved. The formula is as follows:

$$\%\ Families\ Receiving\ Warnings\ in\ Disaster-Prone\ Areas$$
$$= \frac{\sum \text{Families Receiving Warnings in Disaster} - \text{Prone Areas}}{\sum \text{Family living in a disaster} - \text{prone area}} \times 100\%$$

Table 16.2 Disaster risk level classification (Exposed Lives per Type of Disaster per square kilometre)

No.	Parameter	Value	Class	Weight (%)	Score (Class*Weight)
1.	Earthquake	<500	1	40	0.4
		500–1000	2		0.8
		>1000	3		1.2
2.	Tsunami	<500	1	40	0.4
		500–1000	2		0.8
		>1000	3		1.2
3.	Volcanic eruption	<500	1	40	0.4
		500–1000	2		0.8
		>1000	3		1.2
4.	Flood	<500	1	40	0.4
		500–1000	2		0.8
		>1000	3		1.2
5.	Landslide	<500	1	40	0.4
		500–1000	2		0.8
		>1000	3		1.2
6.	Drought	<500	1	40	0.4
		500–1000	2		0.8
		>1000	3		1.2
7.	Land and forest fires	<500	1	30	0.3
		500–1000	2		0.6
		>1000	3		0.9
8.	Extreme weather	<500	1	40	0.4
		500–1000	2		0.8
		>1000	3		1.2
9.	Tidal waves/abrasion	<500	1	40	0.4
		500–1000	2		0.8
		>1000	3		1.2

The next target is families receiving warnings in disaster-prone areas. The village indicators used are the classification of disaster-prone villages, village regulations/village head decrees on disaster mitigation, thoughts for disaster mitigation, mitigation implementation, and disaster warnings from the village government. The hamlets (Rukun Tetangga/RT) indicator used is the existence of disaster management training. The family/individual indicators used are participation in disaster management training. Efforts that can be made to achieve this target include: conducting community preparedness training in facing disasters, conducting volunteer training for disaster management, conducting training in recognising disaster potential and

Table 16.3 Disaster risk level classification (Loss per Type of Disaster per Billion Rupiah)

No.	Parameter	Value	Class	Weight (%)	Score (Class*Weight)
1.	Earthquake	<1.55	1	60	0.6
		1.55–3.30	2		1.2
		>3.30	3		1.8
2.	Tsunami	<1.55	1	50	0.5
		1.55–3.30	2		1
		>3.30	3		1.5
3.	Volcanic eruption	<1.55	1	50	0.5
		1.55–3.30	2		1
		>3.30	3		1.5
4.	Flood	<1.55	1	50	0.5
		1.55–3.30	2		1
		>3.30	3		1.5
5.	Landslide	<1.55	1	50	0.5
		1.55–3.30	2		1
		>3.30	3		1.5
6.	Drought	<1.55	1	30	0.3
		1.55–3.30	2		0.6
		>3.30	3		0.9
7.	Land and forest fires	<1.55	1	30	0.3
		1.55–3.30	2		0.6
		>3.30	3		0.9
8.	Extreme weather	<1.55	1	60	0.6
		1.55–3.30	2		1.2
		>3.30	3		1.8
9.	Tidal wave/abrasion	<1.55	1	50	0.5
		1.55–3.30	2		1
		>3.30	3		1.5

mitigation, and conducting evacuation drills during disasters or extreme climate change incidents.

16.3.10 Early Warning System for Weather and Climate and Disaster Management

The target of the early warning system for weather, climate and disaster management is that the system is expected to be available. The village indicators used are the

Table 16.4 Classification of disaster risk levels (Environmental Damage per Type of Disaster per Hectare)

No.	Parameter	Value	Class	Weight (%)	Score (Class*Weight)
1.	Earthquake	0	1	0	0
		0	2		0
		0	3		0
2.	Tsunami	<55	1	10	0.1
		55–155	2		0.2
		>155	3		0.3
3.	Volcanic eruption	<65	1	10	0.1
		65–185	2		0.2
		>185	3		0.3
4.	Flood	<70	1	10	0.1
		70–205	2		0.2
		>205	3		0.3
5.	Landslide	<65	1	10	0.1
		65–185	2		0.2
		>185	3		0.3
6.	Drought	<65	1	30	0.3
		65–185	2		0.6
		>185	3		0.9
7.	Land and forest fires	<65	1	40	0.4
		65–185	2		0.8
		>185	3		1.2
8.	Extreme weather	0	1	0	0
		0	2		0
		0	3		0
9.	Tidal waves/erosion	<70	1	10	0.1
		70–205	2		0.2
		>205	3		0.3

Table 16.5 Disaster risk level classification (Capacity)

No.	Parameter	Value	Class	Weight	Score (Class*Weight)
1.	Village Capacity	<55	1	100%	1
		55–85	2		2
		>85	3		3

Table 16.6 Classification of disaster risk levels (Risk Index)

No.	Para meter	Value	Capacity=1	Capacity=2	Capacity=3
1.	Earthquake	Low	4	2	1.33
		Medium	16	8	5.33
		High	36	18	12
2.	Tsunami	Low	4	2	1.33
		Medium	16	8	5.33
		High	36	18	12
3.	Volcanic eruption	Low	3	1.5	1
		Medium	12	6	4
		High	27	13.5	9
4.	Flood	Low	4	2	1.33
		Medium	16	8	5.33
		High	36	18	12
5.	Landslide	Low	5	2.5	1.67
		Medium	20	10	6.67
		High	45	22.5	15
6.	Drought	Low	4	2	1.33
		Medium	16	8	5.33
		High	36	18	12
7.	Land and forest fires	Low	4	2	1.33
		Medium	16	8	5.33
		High	36	18	12
8.	Extreme weather	Low	4	2	1.33
		Medium	16	8	5.33
		High	36	18	12
9.	Wave	Low	4	2	1.33
	Surge/abrasion	Medium	16	8	5.33
		Height	36	18	12

classification of disaster-prone villages, village regulations/village head decrees on disaster mitigation, thoughts for disaster mitigation, mitigation implementation, and disaster warnings from the village government. The hamlets (Rukun Tetangga/RT) indicators used are the existence of disaster detection and warning tools. Therefore, all villages need to build an early warning system in disaster-prone centre locations.

16.3.11 Household Waste Management

To measure the presence of household waste management facilities such as the presence of a bin for each family, the number of waste disposal amenities per family is divided by the number of families, then multiplied by 100%. If the result is 100%, then the SDGs Desa target is achieved. The formula is as follows:

$$\%\ Household\ Waste\ Management = \frac{\sum \text{presence of a bin for each family}}{\sum \text{Families}} \times 100\%$$

The target for household waste management is set at 100%. The indicators used are the presence of household waste management facilities and household liquid waste management facilities. Efforts that can be made as part of household waste management include providing bins for organic and inorganic waste, replacing plastic bin liners with newspaper or cardboard, turning organic waste into compost, recycling dry inorganic waste, giving leftover cooking oil to oil-processing agencies, managing hazardous waste, and other actions to reduce plastic waste.

16.3.12 Waste Management Facilities in the Village

To measure the presence of waste management facilities in the village, data on the number of waste/recycling facilities in the village is needed. If there are such facilities, then the SDGs Desa target is achieved.

The target for waste management facilities in the village is set to be present. The hamlets (Rukun Tetangga/RT) indicators used are the number of waste management businesses, the types of waste managed, the number of wastewater management institutions, and the types of transportation in the village that are used and do not cause pollution. Efforts that can be made to build a waste management system for the village area include building a waste bank, implementing a zero-waste programme for household waste, implementing learning programmes about waste sorting, reducing the use of items that cause environmental damage, and conducting waste recycling training.

16.3.13 Presence of Parks/Fields in the Village

To measure the presence of open fields, squares, parks, and other open public spaces, data on the number of open public spaces in the village is needed. If there are open public spaces in the village, then the SDGs Desa target is achieved.

The target for parks/fields in the village is set to parks/fields being present. The hamlets (Rukun Tetangga/RT) indicators used are the size of the village park (square

Table 16.7 Operational definitions for safe and comfortable village residential area

	Keyword	Operational definition
1.	Waste bank	A place used to collect sorted waste
2.	Cangkrukan/ jagongan/kongkow	The activity of village neighbours gathering in front of houses or on the side of village roads without prior arrangement
3.	Inclusive village	A village that is friendly towards marginal groups, such as girls, people with disabilities, and others
4.	Disaster Risk Index (IRB)	Determination of disaster risk level
5.	Household waste	Waste generated from domestic production originating from the kitchen, bathroom, laundry and other household industries
6.	Waste management	Creative methods or steps used to manage waste
7.	Disaster prone potential map	A map that depicts the disaster-prone potential in a region
8.	Slum house	A house that is unfit for habitation, characterised by disorganised construction and poor quality of the building and facilities that do not meet requirements
9.	Environmental Security System (Sistem Kamanan Lingkungan/ Siskamling)	Joint efforts in improving community security and order by providing protection and security for the community, prioritising prevention efforts and countering various forms of threats and disturbances to public order
10.	Early Warning System	A system to notify of the occurrence of natural events, which can be disasters or other natural signs. Early warning to the community about disasters is an action to provide information in a language that is easily understood by the community
11.	Waste disposal site	The place before waste is transported to recycling, management or final place (Tempat Pembuangan Akhir/TPA)
12.	Zero waste	A programme to reuse products that can still be useds

metres) and the size of the village field (square metres). Efforts that can be made to implement greening in the village include: planting trees in the village environment, such as on vacant village land, in home yards, in areas around the watershed, and so on, and creating green open spaces in the village, such as outdoor playgrounds for the community.

16.4 Operational Definitions for Safe and Comfortable Village Residential Area

Key words related to the Safe and Comfortable Village Residential Area are presented in Table 16.7.

References

BNPB. 2020. *IRBI Indonesian Disaster Risk Index 2020* [internet]. Available at: https://inarisk.bnpb.go.id/pdf/BUKU%20IRBI%202020%20KP.pdf.

BPS. 2019. *Percentage of Urban Slum Households (40% and Below), by Province (Percent), 2015-2016* [internet]. Available at: https://sirusa.bps.go.id/sirusa/index.php/indicator/1441.

KemenPUPR. 2020. Strategic Plan of the Directorate of Public and Commercial Housing 2020-2024. Jakarta (ID): KemenPUPR.

Chapter 17
SDGs Desa 12: Environmentally Conscious Village: Consumption and Production

17.1 Concept of Environmentally Conscious Village: Consumption and Production

The SDG Desa Environmentally Conscious Village: Consumption and Production means the use of services and products to meet basic needs and bring a better quality of life for village residents. This is carried out while minimising the use of natural resources and toxic materials, as well as waste emissions and pollutants, so as not to endanger the needs of future generations (UN Environment Programme n.d.). This goal is related to efforts to reduce the environmental impact caused to the earth through reasonable production and consumption patterns. Economic growth is an important indicator in realising the welfare of citizens. However, the economic growth that is created must consider sustainability. Therefore, steps are needed to reduce the ecological footprint by changing the way food and other resources are produced and consumed. Separating economic growth from resource use and environmental degradation is one of the goals of the SDGs Desa.

Every year, it is estimated that a third of all food produced (equivalent to 1.3 billion tons worth about US$1 trillion) ends up rotting in the bins of consumers and retailers or is damaged due to poor transportation and harvesting practices. If the global population reaches 9.6 billion by 2050, the equivalent of nearly three planets will be needed to provide the natural resources needed to maintain the current lifestyle. If people around the world switch to energy-saving light bulbs, the world would save US$120 billion per year (UN n.d.).

Indicators of the success of the SDGs Desa targets can be seen from village policies that regulate waste management in the business world, the occurrence of efficiency in the use of natural resources, and efforts to manage household waste and business waste.

A. H. Iskandar, *The Methods of Measuring the National Sustainable Development Goals in Indonesia*, Engaging Indonesia, https://doi.org/10.1007/978-981-96-0927-7_17

17.2 Targets for Environmentally Conscious Village: Consumption and Production

There are two targets for SDGs Desa 12 Environmentally Conscious Village: Consumption and Production, to be achieved by 2030, namely:

(1) The establishment of village regulations/decisions by the village head about business activities that do not cause pollution, waste management, and household waste
(2) Availability of waste-processing units.

17.3 Measurement of Environmentally Conscious Village: Consumption and Production

17.3.1 Village Regulations/Decisions on Business Activities that Do Not Cause Pollution, Waste Management and Household Waste

To measure this target, data on village regulations, head of village regulations, or decisions by the village head about business activities that do not cause pollution, waste management and household waste are needed. If there is a policy that regulates these, then the SDGs Desa target is achieved.

17.3.2 Waste-Processing Unit

To measure the waste-processing unit, data on the waste-processing unit in the village is needed. If there is one, then the SDGs Desa target is achieved.

17.3.3 Analysis of Environmentally Conscious Village: Consumption and Production

The target for village regulations/decisions on business activities that do not cause pollution, waste management, and household waste is that these are expected to be in place by 2030. The village indicator used is the presence of village regulations/decisions on business activities that do not cause pollution, and the existence of village regulations/head of village decree on household waste management. Efforts that can be made to build a sustainable food production and consumption pattern in

Table 17.1 Operational definitions for environmentally conscious village: consumption and production

	Key word	Operational definition
1.	Food barn	A food reserve institution in rural areas that plays a role in overcoming the community's food vulnerability
2.	Sustainable consumption	A pattern of consumption of goods and services that does not have a negative impact on the environment to meet basic human needs
3.	Sustainable production	A production pattern that ensures ecosystem balance
4.	Environmentally friendly	A product that does not harm the environment

the village include developing village food reserves through village barns, as well as procuring reserves of rice, corn, or sorghum.

The next target is that a waste-processing unit is expected to be available. The Neighbourhood Units/hamlets (Rukun Tetangga/RT) indicator used is the existence of a waste-processing unit and the volume of waste managed. Efforts that can be made to improve village waste management include building a waste bank, implementing a zero-waste programme for household waste, implementing learning programmes about waste sorting and reducing the use of items that cause environmental damage, as well as conducting waste-recycling training. Other actions are to: encourage village communities to create environmentally friendly products through data collection of community business products, develop consultation services related to products to be developed, conduct training for the creation of environmentally friendly products by bringing in experts, and carry out comparative studies for the creation of environmentally friendly and feasible village products.

17.4 Operational Definitions for Environmentally Conscious Village: Consumption and Production

Key words related to Environmentally Conscious Village: Consumption and Production are presented in Table 17.1.

Reference

UN Environment Programme. (n.d.). Sustainable Consumption and Production Policies. [internet]. Available at: https://www.unep.org/explore-topics/resource-efficiency/what-we-do/sustainable-consumption-and-production-policies.

Chapter 18
SDGs Desa 13: Climate Change Responsive Village

18.1 Concept of Climate Change Responsive Village

Climate change means changes in climate caused directly or indirectly by human activities that alter the global atmospheric composition and natural climate variability observed over a comparable period of time (UN 1992). A climate change responsive village is a village that has the necessary disaster risk index as a precautionary measure to minimise disaster victims. With a climate change responsive village, we hope that villages in Indonesia is capable of mitigating the impacts of climate change, including anticipating the disaster impacts it causes.

The impacts of climate change have been felt by almost all countries in the world and have had adverse consequences on life. Currently, there is a climate emergency where the earth is producing new weather patterns, such as rising surface temperatures, changes in planting seasons, changes in rainfall and dry seasons, increasing droughts and heatwaves, rising sea levels, damage to coral reefs, and so on. Although greenhouse gas emissions were projected to fall by about 6% in 2020 due to travel bans and economic slowdowns caused by the COVID-19 pandemic, this increase was only temporary (UN 2020). Climate change does not stop. As the global economy begins to recover from the pandemic, emissions are expected to return to higher levels.

This SDGs Desa aims to help reduce the impact of global climate change, with several programmes that can be carried out by the village government in accordance with its main tasks and functions. The success of achieving this goal can be seen from several indicators, one of which is the disaster risk index in the village.

A. H. Iskandar, *The Methods of Measuring the National Sustainable Development Goals in Indonesia*, Engaging Indonesia, https://doi.org/10.1007/978-981-96-0927-7_18

18.2 Climate Change Responsive Village Targets

To achieve the Climate Change Responsive Village goal, the following three targets have been set to be achieved by 2030:

(1) Disaster Risk Index reaches 0
(2) Disaster mitigation in all Neighbourhood Units/hamlet (Rukun Tetangga/RT)
(3) Creation, maintenance, and normalisation of rivers, canals, and embankments.

18.3 Measurement of Climate Change Responsive Village

18.3.1 Disaster Risk Index

The measurement of the Disaster Risk Index refers to the measurement carried out by the National Disaster Management Agency (BNPB 2020). The data used are (1) hazard per type of disaster, (2) lives exposed per type of disaster, (3) rupiah losses per type of disaster, (4) environmental damage (ha) per type of disaster, and (5) local government capacity per district/city. The determination of the hazard weight per type of disaster is based on the relationship between the frequency of occurrence and the absence of a warning. Table 18.1 shows the hazard weight per type of disaster and the hazard level value: Class 1 for low, Class 2 for medium, and Class 3 for high.

Table 18.2 determines the score based on the exposed lives per type of disaster with an index value of: Class 1 for low, Class 2 for medium, and Class 3 for high.

Table 18.3 determines the score based on the loss in rupiah per type of disaster with an index value of: Class 1 for low, Class 2 for medium, and Class 3 for high.

Table 18.4 determines the score based on environmental damage (ha) per type of disaster with an index value of: Class 1 for low, Class 2 for medium, and Class 3 for high.

Table 18.5 determines the score based on the village's capacity with an index value of: Class 1 for low, Class 2 for medium, and Class 3 for high.

The village can measure the Disaster Risk Index based on the five tables above. The Disaster Risk Index is measured per disaster using the following formula.

The results of each disaster show the level of disaster risk based on Table 18.6. Then, the overall Disaster Risk Index can be obtained by adding up all the Disaster Risk Indexes from each disaster that has been previously calculated.

The Disaster Risk Index target is 0 for all RTs. The village indicators used are the classification of disaster-prone villages, the existence of village regulations/village head decrees on disaster mitigation, thoughts on disaster mitigation, implementation of mitigation, and disaster warnings from the village government. The Neighbourhood Units/hamlet (Rukun Tetangga/RTs) indicators used are the number of disasters, the existence of disaster management training, the number of disaster response training activities in the past year, the origin of disaster management training (government or private), and the existence of early disaster detection tools. Efforts that can

Table 18.1 Disaster risk level classification (hazard per type of disaster)

No.	Parameter	Value (%)	Class	Weight	Score (class * weight)
1	Earthquake	100	1	4	4
			2		8
			3		12
2	Tsunami	100	1	4	4
			2		8
			3		12
3	Volcanic eruption	100	1	3	3
			2		6
			3		9
4	Flood	100	1	4	4
			2		8
			3		12
5	Landslide	100	1	5	5
			2		10
			3		15
6	Drought	100	1	4	4
			2		8
			3		12
7	Land and forest fires	100	1	4	4
			2		8
			3		12
8	Extreme weather	100	1	4	4
			2		8
			3		12
9	Tidal waves/abrasion	100	1	4	4
			2		8
			3		12

be made to maintain and control the impact of climate change include: building bore wells/pumps at fire-prone points; creating/improving ditches in flood-prone areas; procuring flood control equipment; creating slopes and coastal abrasion protection buildings; creating breakwater embankments; buying seeds and planting mangroves; planting on slopes or with landslide retaining concrete structures (retaining walls); procuring waste transport equipment and temporary waste disposal sites; procuring equipment for waste/utilisation (for example, making organic fertiliser, shredding machines); carrying out rehabilitation/relocation of settlements in landslide-prone areas; procuring disaster management support equipment, such as evacuation signs and community-based early warning systems; compiling climate change

Table 18.2 Classification of disaster risk level (exposed lives per type of disaster per square kilometre)

No.	Parameter	Value	Class	Weight (%)	Score (class * weight)
1	Earthquake	<500	1	40	0.4
		500–1000	2		0.8
		>1000	3		1.2
2	Tsunami	<500	1	40	0.4
		500–1000	2		0.8
		>1000	3		1.2
3	Volcanic eruption	<500	1	40	0.4
		500–1000	2		0.8
		>1000	3		1.2
4	Flood	<500	1	40	0.4
		500–1000	2		0.8
		>1000	3		1.2
5	Landslide	<500	1	40	0.4
		500–1000	2		0.8
		>1000	3		1.2
6	Drought	<500	1	40	0.4
		500–1000	2		0.8
		>1000	3		1.2
7	Land and forest fires	<500	1	30	0.3
		500–1000	2		0.6
		>1000	3		0.9
8	Extreme weather	<500	1	40	0.4
		500–1000	2		0.8
		>1000	3		1.2
9	Tidal waves/abrasion	<500	1	40	0.4
		500–1000	2		0.8
		>1000	3		1.2

adaptation and mitigation plans; conducting ProKlim community group training; compiling disaster response plans; conducting socialisation of land clearing without burning; conducting disaster volunteer training; conducting disaster socialisation and simulations; and conducting independent waste-management training.

Table 18.3 Classification of disaster risk levels (loss per type of disaster per billion rupiah

No.	Parameter	Value	Class	Weight (%)	Score (class * weight)
1	Earthquake	<1.55	1	60	0.6
		1.55–3.30	2		1.2
		>3.30	3		1.8
2	Tsunami	<1.55	1	50	0.5
		1.55–3.30	2		1
		>3.30	3		1.5
3	Volcanic eruption	<1.55	1	50	0.5
		1.55–3.30	2		1
		>3.30	3		1.5
4	Flood	<1.55	1	50	0.5
		1.55–3.30	2		1
		>3.30	3		1.5
5	Landslide	<1.55	1	50	0.5
		1.55–3.30	2		1
		>3.30	3		1.5
6	Drought	<1.55	1	30	0.3
		1.55–3.30	2		0.6
		>3.30	3		0.9
7	Land and forest fires	<1.55	1	30	0.3
		1.55–3.30	2		0.6
		>3.30	3		0.9
8	Extreme weather	<1.55	1	60	0.6
		1.55–3.30	2		1.2
		>3.30	3		1.8
9	Tidal waves/erosion	<1.55	1	50	0.5
		1.55–3.30	2		1
		>3.30	3		1.5

18.3.2 Disaster Mitigation

To measure the percentage of disaster mitigation, data on facilities/efforts in each Neighbourhood Units/hamlet (Rukun Tetangga/RTs) (such as rubber boats, tents, masks, etc.) are needed. If the hamlet (Rukun Tetangga/RT) has safety equipment, it is counted as 1 point. The total points are divided by the number of RTs and then multiplied by 100%. The formula is as follows.

$$\%\ \text{Disaster mitigation} = \frac{\sum \text{Safety equipment in hamlet (RT)} = 1}{\sum \text{Neighbourhood Units (RT)}} \times 100\%$$

Table 18.4 Classification of disaster risk levels (environmental damage per type of disaster per hectare)

No.	Parameter	Value	Class	Weight (%)	Score (class * weight)
1	Earthquake	0	1	0	0
		0	2		0
		0	3		0
2	Tsunami	<55	1	10	0.1
		55–155	2		0.2
		>155	3		0.3
3	Volcanic eruption	<65	1	10	0.1
		65–185	2		0.2
		>185	3		0.3
4	Flood	<70	1	10	0.1
		70–205	2		0.2
		>205	3		0.3
5	Landslide	<65	1	10	0.1
		65–185	2		0.2
		>185	3		0.3
6	Drought	<65	1	30	0.3
		65–185	2		0.6
		>185	3		0.9
7	Land and forest fires	<65	1	40	0.4
		65–185	2		0.8
		>185	3		1.2
8	Extreme weather	0	1	0	0
		0	2		0
		0	3		0
9	Tidal waves/erosion	<70	1	10	0.1
		70–205	2		0.2
		>205	3		0.3

Table 18.5 Disaster risk level classification (capacity)

No.	Parameter	Value	Class	Weight (%)	Score (class * weight)
1	Village capacity	<55	1	100	1
		55–85	2		2
		>85	3		3

Table 18.6 Classification of disaster risk levels (risk index)

No.	Parameter	Value	Capacity = 1	Capacity = 2	Capacity = 3
1	Earthquake	Low	4	2	1.33
		Medium	16	8	5.33
		High	36	18	12
2	Tsunami	Low	4	2	1.33
		Medium	16	8	5.33
		High	36	18	12
3	Volcanic eruption	Low	3	1.5	1
		Medium	12	6	4
		High	27	13.5	9
4	Flood	Low	4	2	1.33
		Moderate	16	8	5.33
		High	36	18	12
5	Landslide	Low	5	2.5	1.67
		Moderate	20	10	6.67
		High	45	22.5	15
6	Drought	Low	4	2	1.33
		Moderate	16	8	5.33
		High	36	18	12
7	Land and forest fires	Low	4	2	1.33
		Moderate	16	8	5.33
		High	36	18	12
8	Extreme weather	Low	4	2	1.33
		Moderate	16	8	5.33
		High	36	18	12
9	Wave tide/abrasion	Low	4	2	1.33
		Medium	16	8	5.33
		High	36	18	12

18.3.3 Creation, Maintenance, and Normalisation of Rivers, Canals, and Levees

To measure this target, data on the creation, maintenance, and normalisation of rivers, canals, and levees in each Neighbourhood Units/hamlet (Rukun Tetangga/RTs) are needed. If the hamlet (Rukun Tetangga/RT) carry out the creation, maintenance, and normalisation of rivers, canals, and levees, then it is counted as one point. The total points are then divided by the number of RTs and multiplied by 100%. The formula is as follows.

Table 18.7 Operational definitions for climate change responsive village

	Key word	Operational definition
1	Mitigation	Efforts to reduce disaster risk
2	Climate change	The condition when changes occur in the earth's climate system resulting in new weather patterns, for example, rising surface temperatures, changes in planting seasons, changes in rainfall and dry seasons, increasing droughts and heatwaves, polar ice melting and making the poles ice-free, rising sea levels, damage to coral reefs, and so on

$$\begin{aligned}&\%\ \text{Existence of creation, maintenance, and}\\&\text{normalisation of rivers, canals, and levees}\\&= \frac{\sum \begin{matrix}\text{Creation, maintenance, and normalisation of}\\ \text{rivers, canals, and levees in hamlet (RT)} = 1\end{matrix}}{\sum \text{Neighbourhood Units (RT)}} \times 100\%\end{aligned}$$

18.4 Operational Definitions for Climate Change Responsive Village

Key words related to Climate Change Responsive Village are presented in Table 18.7.

References

BNPB. 2020. *IRBI Indonesian disaster risk index 2020* [internet]. https://inarisk.bnpb.go.id/pdf/BUKU%20IRBI%202020%20KP.pdf
UN. 1992. *United Nations framework convention on climate change.* [internet]. https://unfccc.int/files/essential_background/background_publications_htmlpdf/application/pdf/conveng.pdf
UN. 2020. *Fall in COVID-linked carbon emissions won't halt climate change—UN weather agency chief* [internet]. https://news.un.org/en/story/2020/04/1062332

Chapter 19
SDGs Desa 14: Village Caring for the Marine Environment

19.1 Concept of Village Caring for the Marine Environment

The goal of this SDGs Desa is to protect the coast and the sea. Caring for the marine environment or 'Life Below Water' in the global SDGs means maintaining the marine ecosystem by conserving and sustainably utilising marine, oceanic and maritime resources for sustainable development (UN n.d.).

More than 3 billion people depend on the sea and coastal biodiversity for their livelihoods. However, 30% of the world's fish stocks are overexploited. Marine pollution, mostly originating from land, has reached worrying levels, with an average of 13,000 pieces of plastic waste found in every square kilometre of ocean (Bappenas n.d.).

SDGs Desa number 14 aims to regulate and protect marine and coastal ecosystems and strengthen the protection and use of sustainable marine resources. To measure the success of achieving this goal, several indicators are used including village policy related to the protection of marine resources, an increase in reasonable fish catches, and no illegal fishing.

19.2 Targets for Village Caring for the Marine Environment

To achieve the goal of the Village Caring for the Marine Environment, the following five targets have been set to be achieved by 2030:

(1) Availability of village regulations/head of village decree on marine resource protection

A. H. Iskandar, *The Methods of Measuring the National Sustainable Development Goals in Indonesia*, Engaging Indonesia, https://doi.org/10.1007/978-981-96-0927-7_19

(2) Availability of village regulations/head of village decree on village spatial planning
(3) Increasing volume of reasonable fish catch according to the type of fish
(4) The area of water conservation areas is at least 33% of the village area
(5) No illegal fishing.

19.3 Measurement of Targets for Village Caring for the Marine Environment

19.3.1 Village Regulations/Head of Village Decree on Marine Resource Protection

The first target is to have in place village regulations/head of village decree on marine resource protection. The village indicator used is the existence of village policies such as village regulations/head of village decree on marine resource protection. Efforts that can be made to protect, preserve, and utilise water resources include rehabilitating mangroves, planting mangroves, cleaning the area around the beach (beach clean-up), installing environmental preservation information boards, providing education/training on coastal resource conservation, conducting coral reef rehabilitation training, socialising the use of environmentally friendly fishing gear, and developing Tourism Villages based on Marine Ecotourism.

To measure the existence of village regulations/head of village decree on marine resource protection, village regulations data, head of village regulation data, and head of village decree data on marine resource protection are needed. If marine resource protection is already controlled in the regulations, then the SDGs Desa target is achieved.

19.3.2 Village Regulations/Head of Village Decree on Village Spatial Planning

The second target is to have in place village regulations/head of village decree on village spatial planning. The village indicator used is the existence of village policies such as village regulations/head of village decree on village spatial planning. Village spatial planning that can be done includes allocating utilisation space for seaweed cultivation and carrying out fish catch utilisation activities in the waters around the village.

To measure the existence of village regulations/head of village decree on village spatial planning, village regulations data, head of village regulation data, and head of village decree data on village spatial planning are needed. If village spatial planning is already controlled in the regulations, then the SDGs Desa target is achieved.

19.3.3 Increasing Volume of Reasonable Fish Catch According to Type of Fish

The volume of reasonable fish catch according to the type of fish is targeted to increase. The head of household/individual indicators used are the type of fish caught, the number of fish caught in the last month (kg), and the number of fish caught in the previous month (kg). Efforts that can be made to regulate reasonable fish catch include conducting regular data collection of fishermen's catch (at least monthly), and socialising the impact of excessive and habitat-damaging fish catch.

Increasing the volume of reasonable fish catch according to the type of fish is based on potential estimates, the allowable catch amount, and the level of utilisation of fish resources in the Management Area of the Republic of Indonesia's Fisheries (Wilayah Pengelolaan Perikanan Negara Republik Indonesia/WPPNRI) (see Table 19.1).

19.3.4 Area of Village Water Conservation

The target area for water conservation is set at a minimum of 33% of the village area. The Neighbourhood Units/hamlets (Rukun Tetangga/RT) indicator used is the area of water conservation (m^2) in the hamlets (Rukun Tetangga/RT). Efforts that can be made for water conservation areas, for example, involve the community especially the younger generation in supervising and managing water conservation areas.

To measure the percentage of the village water conservation area, the village water conservation area is divided by the total area of the village then multiplied by 100%. If the result is more than 33%, then the SDGs Desa target is achieved. The formula is as follows:

$$\%\text{ The village water conservation area} = \frac{\text{The village water conservation area}}{\text{Total area of the village}} \times 100\%$$

19.3.5 No Illegal Fishing

The target for illegal fishing is none. The Neighbourhood Units/hamlets (Rukun TetanggaRT) indicator used is the number of illegal fishing incidents from the last year and the number of illegal fishing incidents from the previous year. Efforts that can be made to suppress cases of illegal fishing and other damaging marine ecosystem actions include: cooperation and coordination of illegal fishing prevention with law enforcement officers, providing assistance with environmentally friendly fishing

Table 19.1 Estimates of potential, allowable catch amount and level of utilisation of fish resources in Management Area of the Republic of Indonesia's Fisheries

Republic of Indonesia's fisheries management area			Small pelagic fish	Large pelagic fish[a]	Demersal fish	Reef fish	Penaeid shrimp	Lobster	Crab	Blue swimming crab	Squid	Total
Malacca Strait and Andaman Sea	WPPNRI 571	Potential (tons)	99,865	64,444	145,495	20,030	59,455	673	12,829	13,614	9,038	425,444
		JTB (tons)	79,892	51,556	116,396	16,024	47,564	539	10,263	10,891	7,230	–
		Utilization rate	0.83	0.52	0.33	0.34	1.59	1.30	1.00	0.93	0.62	–
Indian Ocean west of Sumatra and the Sunda Strait	WPPNRI 572	Potential (tons)	527,029	276,755	362,005	40,570	8,023	1,483	9,543	989	14,579	1,240,975
		JTB (tons)	421,623	221,404	289,604	32,456	6,418	1,186	7,634	791	11,663	–
		Utilization rate	0.50	0.95	0.57	0.33	1.53	0.93	0.18	0.49	0.39	–
The Indian Ocean south of Java to the south of Nusa Tenggara, Sawu Sea and the western part of the Timor Sea	WPPNRI 573	Potential (tons)	630,521	586.128	7,902	22,045	7,340	970	526	3,913	8,195	1,267,540
		JTB (tons)	504,417	468,902	6,322	17,636	5,872	776	421	3,130	6,556	–
		Utilization rate	1.50	1.06	0.39	1.09	1.70	0.61	0.28	0.98	1.11	–
Karimata Strait Waters, Natuna Sea, and South China Sea	WPPNRI 711	Potential (tons)	330,284	185,855	131,070	20,625	62,342	1,421	2,318	9,711	23,499	767.126
		JTB (tons)	264,227	148,684	104,856	16,500	49,873	1,137	1,854	7,769	18,799	–

(continued)

Table 19.1 (continued)

Republic of Indonesia's fisheries management area			Small pelagic fish	Large pelagic fish[a]	Demersal fish	Reef fish	Penaeid shrimp	Lobster	Crab	Blue swimming crab	Squid	Total
		Utilization rate	1.41	0.93	0.61	1.53	0.53	0.54	1.09	1.18	1.84	–
Java Sea Waters	WPPNRI 712	Potential (tons)	364,663	72,812	657,525	29,951	57,965	989	7,664	23,508	126,554	1,341,632
		JTB (tons)	291,730	58,250	526.020	23,961	46,372	791	6.131	18,806	101.24 4	–
		Utilization rate	0.38	0.63	0.83	1.22	1.11	1.36	0.70	0.65	2.02	–
Makassar Strait Waters, Bone Bay, Flores Sea, and Bali Sea	WPPNRI 713	Potential (tons)	208,414	645,058	252,869	19,856	30,404	927	4,347	5,463	10,519	1,177,857
		JTB (tons)	166,731	516,046	202,295	15,885	24,324	742	3,477	4,370	8,415	–
		Utilization rate	1.23	1.13	0.96	1.27	0.52	1.40	0.83	0.73	1.19	–
Tolo Bay Waters and Banda Sea	WPPNRI 714	Potential (tons)	165,944	304.293	98,010	145,530	3,180	724	1,145	1,669	68,444	788,939
		JTB (tons)	132,755	243,435	78,408	116,424	2,544	579	916	1,335	54,755	–
		Utilization rate	0.44	0.78	0.58	0.76	0.39	1.73	1.55	0.77	1.00	–
Tomini Bay Waters, Maluku Sea, Halmahera Sea, Seram Sea, and Berau Bay	WPPNRI 715	Potential (tons)	555,982	31,659	325,080	310,866	6,436	846	891	495	10,272	1,242,526
		JTB (tons)	444,786	25,327	260,064	248,693	5,149	677	712	396	8,217	–
		Utilization rate	0.88	0.97	0.22	0.34	0.78	1.32	1.19	0.98	1.86	–

(continued)

Table 19.1 (continued)

Republic of Indonesia's fisheries management area			Small pelagic fish	Large pelagic fish[a]	Demersal fish	Reef fish	Penaeid shrimp	Lobster	Crab	Blue swimming crab	Squid	Total
Sulawesi Sea waters and the north of Halmahera Island	WPPNRI 716	Potential (tons)	332,635	181,491	36,142	34,440	7,945	894	2,196	294	1,103	597,139
		JTB (tons)	266.108	145,193	28,914	27,552	6,356	715	1,756	235	883	–
		Utilization rate	0.48	0.63	0.45	1.45	0.50	0.75	0.38	0.50	1.42	–
The waters of Cendrawasih Bay and the Pacific Ocean	WPPNRI 717	Potential (tons)	829,188	65,935	131,675	15,016	9,150	1,044	489	58	2,140	1,054,695
		JTB (tons)	663,350	52,748	105,340	12,013	7,320	835	391	46	1,712	–
		Utilization rate	0.70	1.00	0.39	0.91	0.46	1.04	0.87	1.21	1.09	–
Aru Sea, Arafuru Sea and East Timor Sea	WPPNRI 718	Potential (tons)	836,973	818,870	876,722	29,485	62,842	1,187	1,498	775	9,212	2,637,565
		JTB (tons)	669,579	655,096	701,378	23,588	50,274	950	1,198	620	7,370	–
		Utilization rate	0.51	0.99	0.67	1.07	0.86	0.97	0.85	0.77	1.28	–
Total potential (tonnes)												12,541,438

[a] Non-Tuna-Cakalang large pelagic fish

Explanation of Utilisation rate (E)

$E < 0.5$ = Moderate, fishing efforts can be increased

$0.5 \leq E < 1$ = Fully-exploited, fishing efforts should be maintained with strict monitoring

$E \geq 1$ = Over-exploited, fishing efforts should be reduced

Source Decision of the Minister of Maritime Affairs and Fisheries of the Republic of Indonesia Number 50/KEPMEN-KP/2017 on the Estimation of Potential, Allowable Catch, and Utilisation Rate of Fish Resources in the Fisheries Management Area of the Republic of Indonesia

Table 19.2 Operational definitions for village caring for the marine environment

	Key word	Operational definition
1	Illegal fishing	Fishing that is carried out in violation of the laws established in a country's waters
2	Coastal village	A village located on the border between land and sea and that has direct access to the coastal ecosystem (sandy or rocky), mangroves, estuaries, seagrass beds, and coral reef ecosystems
3	Marine ecotourism	Environmental tourism (eco-tourism) based on marine attractions in a location or area dominated by water or marine. Marine ecotourism presents a unique marine ecosystem such as mangrove forests, marine parks, and various fauna, both marine fauna and fauna around the beach
4	Marine resources	Resources that include a wide scope covering marine life (flora and fauna, ranging from microscopic organisms to killer whales and marine habitats) from deep waters to the tidal area on the highland coast and a wide estuary area

gear, providing counselling on the impact of illegal fishing, as well as registering individuals or communities who commit illegal fishing.

To measure the presence or absence of illegal fishing, the number of illegal fishing incidents that occur in coastal villages is needed. If there are none, or there is no longer any illegal fishing, then the SDGs Desa target is achieved.

19.4 Operational Definitions for Village Caring for the Marine Environment

Key words related to Village Caring for the Marine Environment are presented in Table 19.2.

References

Bappenas. n.d.(a). *15 Terrestrial ecosystems protecting, restoring and enhancing sustainable use of terrestrial ecosystems, managing forests sustainably, stopping desertification, recovering land degradation, and stopping loss of biodiversity* [internet]. http://sdgs.bappenas.go.id/tujuan-15/

Bappenas. n.d.(b). *14 Ocean ecosystems preserve and sustainably utilise marine and ocean resources for sustainable development* [internet]. http://sdgs.bappenas.go.id/goal-14/

UN. n.d.(a). *Goal 12: Ensure sustainable consumption and production patterns* [internet].

UN. n.d.(b). *Goal 15* [internet]. https://sdgs.un.org/goals/goal15

UN. n.d.(c). *Goal 16: Promote just, peaceful and inclusive societies* [internet]. https://www.un.org/sustainabledevelopment/peace-justice/

UN. n.d.(d). *Life below water: Why it matters* [internet]. https://www.un.org/sustainabledevelopment/wp-content/uploads/2019/07/14_Why-It-Matters-2020.pdf

Chapter 20
SDGs Desa 15: Village Caring for the Land Environment

20.1 Concept of Village Caring for the Land Environment

Caring for the land environment is an effort to protect, restore and increase the sustainable use of terrestrial ecosystems, manage forests sustainably, stop desertification, restore degraded land, and stopping the loss of biodiversity (UN n.d.). The need for food continues to increase globally. To meet this, plants that provide 80% of food materials are needed. Forests cover 30% of the earth's surface and are habitats for millions of species and are a source of clean water and air and also are very important in combating climate change. Meanwhile, arable land loss reaches 30–35 times the average amount in history. Long droughts and areas of land changing into deserts also increase every year, reaching 12 million hectares and affecting poor communities around the world. Of the 8,300 known animals, 8% have become extinct and 22% are on the brink of extinction (Bappenas n.d.).

This SDGs Desa strives to protect and improve the use of terrestrial ecosystems such as forests, swamps, land, and mountains. Also, promoting sustainable forest management and stopping deforestation is vital to stop the impact of climate change. For the fulfilment of human needs now and in the future, and for the protection of wildlife, the SDGs Desa sets a goal to protect natural resources and wildlife. To see the success of achieving this goal, indicators are used including village government policies related to biodiversity conservation efforts, the extent of green open land, and the number of endangered species.

20.2 Targets for Village Caring for the Land Environment

To achieve the Village Caring for the Land Environment goal, the following seven targets have been set to be achieved by 2030:

(1) The area of open land planted is at least 33% of the village area

A. H. Iskandar, *The Methods of Measuring the National Sustainable Development Goals in Indonesia*, Engaging Indonesia, https://doi.org/10.1007/978-981-96-0927-7_20

(2) The area of damaged forest land reaches 0% of the forest area
(3) 100% use of wood from maintained/restored forests
(4) Critical land in forests reaches 0% of the forest area
(5) Increase of 50% in the population of endangered species
(6) 100% prosecution of environmental offenders
(7) Availability of Village Regulation/Head of Village Decree on biodiversity conservation.

20.3 Measurement of Village Caring for the Land Environment

20.3.1 Area of Open Land Planted

The target for the area of open land planted is set at a minimum of 33% of the village area. The Neighbourhood Units/hamlet (Rukun Tetangga/RT) indicator used is the area of agricultural and plantation land in the last year and the year before, as well as the area of forest land in the last year and the year before. Efforts that can be made to update village land/forest data are, for example, by conducting a survey of the potential and area of village land/forest.

To measure the percentage of open land planted, the area of open land planted is divided by the total area of the village then multiplied by 100%. If the result is more than 33%, then the SDGs Desa target is achieved. The formula is as follows:

$$\%\text{ Open land planted} = \frac{\text{Open land planted}}{\text{Total area of the village}} \times 100\%$$

20.3.2 Area of Damaged Forest Land

The target for the area of damaged forest land is set at 0% of the forest area. The Neighbourhood Units/hamlet (Rukun Tetangga/RT) indicator used is the area of conservation forest damage over the past year and the area of conservation forest damage the previous year. Efforts that can be made to protect, preserve, and utilise forest resources include conducting a survey of village land/forest damage, reforestation and rehabilitation of forests and peatlands, as well as providing education on the impact of conservation forest damage.

To measure the percentage of damaged forest land area, the area of damaged forest land, for example polluted land, is divided by the area of the forest that exists in the village, then multiplied by 100%. If the result is 0%, then the SDGs Desa target is achieved. The formula is as follows:

$$\% \text{ Damaged forest land area} = \frac{\text{Damaged forest land area}}{\text{The area of the forest that exists in the village}} \times 100\%$$

20.3.3 *Utilisation of Maintained/Restored Wood*

The target for wood utilisation is set at 100% of the maintained/restored forest. The Neighbourhood Units/hamlet (Rukun Tetangga/RT) indicator used is the amount of forest produce from the last year and the amount of forest produce from the previous year. Efforts that can be made to increase wood utilisation, for example, by updating the total forest production data regularly, and imposing sanctions on those who exploit forest products without carrying out restoration.

To measure the percentage of wood or forest product utilisation that is maintained/restored, the area of wood or forest product utilisation that is maintained/restored is divided by the total area of wood or forest product utilisation, then multiplied by 100%. If the result is 100%, then the SDGs Desa target is achieved. The formula is as follows:

$$\% \text{ Wood or forest product utilisation that is maintained/restored} = \frac{\text{The area of wood or forest product utilisation that is maintained/restored}}{\text{Total area of wood or forest product utilisation}} \times 100\%$$

20.3.4 *Critical Land in Forest*

The target for critical land in the forest is set at 0% of the forest area. The Neighbourhood Units/hamlet (Rukun Tetangga/RTs) indicator used is the area of critical land (barren/deforested) over the past year compared to the area of critical land (barren/deforested) the previous year. Efforts that can be made to suppress/reduce the area of critical village land include conducting a survey of critical village land, conducting reforestation, and repairing damaged land through the activity of creating village forests that are managed sustainably.

To measure the percentage of critical land in the forest, the area of critical land in the forest is divided by the total forest area in the village then multiplied by 100%. If the result is 0%, then the SDGs Desa target is achieved. The formula is as follows:

$$\% \text{ The area of critical land in the forest} = \frac{\text{The area of critical land in the forest}}{\text{The total forest area in the village}} \times 100\%$$

20.3.5 *Increase in Population of Endangered Species*

The endangered species are targeted to increase their population by 50%. The Neighbourhood Units/hamlet (Rukun Tetangga/RT) indicator used is the list of endangered species, the number of each endangered species in the last year, and the number of each endangered species in the previous year. Efforts made to prevent and suppress the number of endangered species include conducting data collection of types and numbers of endangered species, cooperating and coordinating with related services/parties to increase the population of endangered species, installing information boards of endangered species, and conducting socialisation of the movement to care for rare species that are almost extinct.

To measure the increase in the population of endangered species, the number of populations of endangered species in a certain year, minus the number of populations of endangered species in the previous year, is then divided by the number of populations of endangered species in the previous year, then multiplied by 100%. If the result is above 50%, then the SDGs Desa target is achieved. The formula is as follows:

$$\% \text{ Increasing population of endangered species} = \frac{\sum \text{Population of endangered pecies in a } \textbf{certain year} - \sum \text{Population of endangered species in the } \textbf{previous year}}{\sum \text{Population of endangered species in the } \textbf{previous year}} \times 100\%$$

20.3.6 *Environmental Offenders Prosecuted*

The target is for 100% of environmental offenders to be prosecuted. The Neighbourhood Units/hamlet (Rukun Tetangga/RT) indicator used is the number of residents convicted for environmental violations in the last year and the number of residents convicted for environmental violations in the previous year. Efforts that can be made to reduce hunting and trading of protected flora and fauna species and other environmental damage actions include: recording cases and residents convicted due to environmental violations, as well as cooperating and coordinating with relevant authorities/parties to prevent an increase in environmental damage actions.

To measure the percentage of convicted environmental offenders, the number of convicted environmental offenders is divided by the total number of environmental offenders then multiplied by 100%. If the result is 100%, then the SDGs Desa target is achieved. The formula is as follows:

$$\% \text{ The convicted environmental offenders} = \frac{\sum \text{The convicted environmental offenders}}{\sum \text{The environmental offenders}} \times 100\%$$

20.3.7 Village Regulation/Head of Village Decree on Biodiversity Conservation

The target is to have in place village regulation or head of village decree on biodiversity conservation. The village indicator used is the existence of a village regulation/head of village decree on biodiversity conservation. Efforts that can be made to create sustainable development are, for example, developing a model of village development planning and budgeting that is sensitive to environmental issues.

To measure the existence of any village regulation/head of village decree on biodiversity conservation, village regulation data, head of village regulation data, and head of village decree data on biodiversity conservation are needed. If biodiversity conservation is already controlled in the regulations, then the SDGs Desa target is achieved.

20.4 Operational Definitions for Village Caring for the Land Environment

Key words related to Village Caring for the Land Environment are presented in Table 20.1.

Table 20.1 Operational definitions for village caring for the land environment

	Key word	Operational definition
1	Village forest	A state forest managed by the village and utilised for the welfare of the village, which has not yet been burdened with permits/rights
2	Biodiversity	The diversity of organisms that shows the whole or totality of variation
3	Conservation	Efforts to preserve the environment
4	Peatland	Land consisting of accumulated semi-decomposed plant residues; therefore, its organic content is high
5	Critical land	Land that has suffered damage, thus losing or reducing its function to a certain or expected limit
6	Plantation land	Large-scale plantation business land, usually located in tropical or subtropical areas, used to produce trade commodities (plantations) on a large scale and marketed to distant places, not for local consumption
7	Agricultural land	Large-scale agricultural business land, usually located in tropical or subtropical areas, used to produce trade commodities (agriculture) on a large scale and marketed to distant places, not for local consumption
8	Green open space	Land whose use is open, serving as a place for plants to grow naturally or intentionally planted, and that has ecological functions
9	Wildlife	The uniqueness and/or diversity of animal species whose survival can be facilitated through habitat management.
10	Restoration	Efforts to restore, recover, repair something to its original condition and form
11	Forest resources	All things originating from the forest that can be used to meet human needs

References

Bappenas. n.d. *15 Terrestrial ecosystems protecting, restoring and enhancing sustainable use of terrestrial ecosystems, managing forests sustainably, stopping desertification, recovering land degradation, and stopping loss of biodiversity* [internet]. http://sdgs.bappenas.go.id/tujuan-15/

UN. n.d. *Goal 15* [internet]. https://sdgs.un.org/goals/goal15

Chapter 21
SDGs Desa 16: Village of Peace and Justice

21.1 Concept of Village of Peace and Justice

Conflict occurs in societies that are constantly moving and changing dynamically, including in village communities. This can happen due to the economic gap, marginalisation and discrimination, and poor village governance leading to social injustice in the village administration process, especially during village development. One of the problem-solving efforts is to realise peace in the village.

A Village of Peace and Justice is a village that facilitates the village community especially in relation to the utilisation of village development resources, for prevention and handling of social conflict in the village. The aim is to create a safe village condition, thus ensuring the village government can work fairly and effectively. SDGs Desa goal 16 enables the village to independently guarantee ease and equality of access to justice for the entire village community, without exception.

Some regions enjoy sustained peace, security and prosperity, while other regions seem to be constantly in a cycle of conflict and violence. High levels of violence and insecurity have a damaging impact on development, affecting economic growth, and often causing resentment among citizens that can last for several generations. Sexual violence, crime, exploitation and torture, also often occur in conflict conditions, or if there is no governing law. The state must take action to protect those at risk. In addition, the birth of about one in four children under the age of five worldwide is never officially recorded, causing them to lose important legal identity proof for the protection of their rights and for access to justice and social services (UN n.d.).

This SDGs Desa goal sets several targets to be achieved by 2030, namely the absence of criminal incidents, fights, domestic violence (DV), and violence against children, as well as the preservation of a culture of mutual cooperation in the village, improving the democracy index in the village, and the absence of human trafficking and child labour.

A. H. Iskandar, *The Methods of Measuring the National Sustainable Development Goals in Indonesia*, Engaging Indonesia, https://doi.org/10.1007/978-981-96-0927-7_21

21.2 Village of Peace and Justice Targets

To achieve the Village of Peace and Justice goal, the following 31 targets are set to be achieved by 2030:

(1) Incidents of violence against children reach 0%
(2) Incidents of fights reach 0%
(3) Incidents of domestic violence (DV) reach 0%
(4) Criminal incidents reach 0%
(5) Mutual cooperation among residents of different religions, races, and groups
(6) Child labour reaches 0%
(7) Human trafficking reaches 0%
(8) Poor people who receive litigation and non-litigation legal aid reach 100%
(9) Judicial services for the poor through court sessions outside the courthouse reach 100%
(10) Case fee waivers apply to 100% of poor people
(11) Legal Service Posts are available
(12) Anti-Corruption Behaviour Index (Indeks Perilaku Anti Korupsi/IPAK) reaches a score of 5
(13) The latest financial report was received in the Village Consultative Meeting
(14) The financial report can be accessed by the public
(15) The latest Village Head's accountability report was received in the Village Consultative Meeting
(16) The procurement process for goods and services for all development projects is open to the public
(17) The Village Government's Organisational Structure is in accordance with applicable laws and regulations
(18) The level of citizen satisfaction with village government services is high
(19) Women in the Village Consultative Agency (Badan Permusyawaratan Desa/ BPD) reach at least 30%
(20) Women in the village government reach at least 30%
(21) The Democracy Institution Index reaches 100
(22) The Civil Liberties Index reaches 100
(23) The Political Rights Index reaches 100
(24) Coverage of birth certificate ownership for the lowest 40% of the income population reaches 100%
(25) Children who have a birth certificate reach 100%
(26) Handling of complaints of violations due to race, ethnicity, and religion reaches 100%
(27) Development planning documents and financial report documents that can be accessed by the public reach 100%
(28) Summary of development planning documents and financial report documents are announced to the public in an open space
(29) Resolution of complaints related to public information reaches 100%
(30) The village government provides information on the same day.

21.3 Measurement of Village of Peace and Justice

21.3.1 Incidents of Violence Against Children

The target for incidents of violence against children is set at 0%. The Neighbourhood Units/hamlet (Rukun Tetangga/RT) indicator used is the number of cases of violence against children in the last year and the number of cases of violence against children in the previous year. Efforts that can be made include conducting socialisation and facilitating prevention and handling of cases of violence against children through the development of child-sensitive planning mechanisms.

To measure the percentage of incidents of violence against children in each Neighbourhood Units/hamlet (Rukun Tetangga/RT), the number of incidents of violence against children in each hamlet (Rukun Tetangga/RT) is divided by the number of hamlet (Rukun Tetangga/RT), then multiplied by 100%. If the result is 0%, then the SDGs Desa target is achieved. The formula is as follows:

$$\begin{aligned}&\%\ \text{Incidents of violence against children}\\&\quad\text{in each Neighbourhood Unit}\\&=\frac{\begin{array}{c}\sum\text{Incidents of violence against children}\\\text{in each Neighbourhood Unit}\end{array}}{\sum\text{Neighbourhood Units (RT)}}\times 100\%\end{aligned}$$

21.3.2 Incidents of Fighting

The target for incidents of fighting is set at 0%. The Neighbourhood Units/hamlet (Rukun Tetangga/RT) indicator used is the incidents of fighting in the last year and incidents of fighting in the previous year. Efforts that can be made to meet the target include, for example, establishing a social service mediation and conflict resolution post.

To measure the percentage of fighting incidents in each Neighbourhood Unit/hamlet (Rukun Tetangga/RT), the number of fighting incidents in each hamlet (Rukun Tetangga/RT) is divided by the number of /hamlet (Rukun Tetangga/RT), then multiplied by 100%. If the result is 0%, then the SDGs Desa target is achieved. The formula is as follows:

$$\begin{aligned}&\%\ \text{Fighting incidents in each Neighbourhood Units}\\&=\frac{\sum\text{Fighting incidents in each Neighbourhood Units}}{\sum\text{Neighbourhood Units (RT)}}\times 100\%\end{aligned}$$

21.3.3 Domestic Violence Incidents (DVI)

The target for domestic violence incidents (DVI) is set at 0%. The Neighbourhood Units/hamlet (Rukun Tetangga/RT) indicator used is the number of DVI cases over the past year and the number of DVI cases from the previous year. The head of household/individual indicator used is the number of family members who are DVI victims. Efforts that can be made include conducting socialisation and facilitating prevention and handling of Domestic Violence Incidents (DVI) through the development of planning mechanisms that are sensitive to women, developing social assistance posts/services for mediation and resolution of social conflicts, training in Preparation for Marriage for Adolescents of Marriageable Age (including conflict management in relationships and households), training and socialisation about prevention and handling of violence against women (including human trafficking crimes), conducting basic education seminars related to peaceful families (including material related to prevention of domestic violence), and the Stop DVI campaign, especially violence against women.

To measure the percentage of domestic violence incidents (DVI) in each Neighbourhood Unit/hamlet (Rukun Tetangga/RT), the number of DVI incidents in each Neighbourhood Unit/hamlet (Rukun Tetangga/RT) is divided by the number of hamlet (Rukun Tetangga/RT), then multiplied by 100%. If the result is 0%, then the SDGs Desa target is achieved. The formula is as follows:

$$\begin{aligned}&\%\ \text{Domestic violence incidents in each Neighbourhood Units}\\&=\frac{\sum \text{Domestic violence incidents in each Neighbourhood Units}}{\sum \text{Neighbourhood Units (RT)}}\times 100\%\end{aligned}$$

21.3.4 Crime Incidents

The target for crime incidents is set at 0%. The Neighbourhood Units/hamlet (Rukun Tetangga/RT) indicator used is the number of crimes in the last year and the number of crimes in the previous year. Efforts that can be made to meet the target include, for example prosecuting the perpetrators of crime.

To measure the percentage of crime incidents in each Neighbourhood Unitshamlet (Rukun Tetangga/RT), data on the number of crime incidents in each hamlet (Rukun Tetangga/RT) is divided by the number of hamlet (Rukun Tetangga/RT), then multiplied by 100%. If the result is 0%, then the SDGs Desa target is achieved. The formula is as follows:

$$\begin{aligned}&\%\ \text{Crime incidents in each Neighbourhood Units}\\&=\frac{\sum \text{Crime incidents in each Neighbourhood Units}}{\sum \text{Neighbourhood Units (RT)}}\times 100\%\end{aligned}$$

21.3.5 Mutual Cooperation among Residents of Different Religions, Races, and Groups

The next goal is to foster mutual cooperation among residents of different religions, races, and groups. The village indicator used is the existence of mutual cooperation among residents of different religions, races, and groups. The head of household/individual indicator used is involvement in mutual cooperation among residents of different religions, races, and groups. Efforts that can be made to increase mutual cooperation among religions, races, and groups include facilitating mutual cooperation activities to clean up the environment, build village infrastructure, and help residents in need.

To measure the implementation of mutual cooperation among residents of different religions, races, groups, the number of residents participating in community service aged over 15 years is divided by the number of residents aged over 15 years, then multiplied by 100%. If there is resident participation in community service activities, then the SDGs Desa is achieved. The formula is as follows:

$$\%\ \text{Implementation of mutual cooperation among residents} = \frac{\sum \text{Residents participating in community service aged 15 years}}{\sum \text{Residents aged 15 years}} \times 100\%$$

21.3.6 Child Labour

The target for child labour is set at 0%. The indicators used are the age of family members and the number of children who work. Efforts that can be made to eliminate child labour include conducting socialisation about the adverse effects of child labour, as well as monitoring the presence of child labour in the village.

To measure the percentage of child labour, the number of children under the age of 15 years who work is divided by the total number of residents who are working, then multiplied by 100%. If the result is 0%, then the SDGs Desa target is achieved. The formula is as follows.

$$\%\ \text{Child labour} = \frac{\sum \text{Working Children aged} < 15\ \text{years}}{\sum \text{Working Village Residents}} \times 100\%$$

21.3.7 Human Trafficking

The target for human trafficking is set at 0%. The indicator used is the number of residents involved in trafficking. Efforts that can be made to eliminate human trafficking include conducting socialisation about the dangers of human trafficking, as well as monitoring the potential for human trafficking in the village.

To measure the percentage of human trafficking in each Neighbourhood Unit/ hamlet (Rukun Tetangga/RT), the number of human trafficking cases in each hamlet (Rukun Tetangga/RT) is divided by the number of hamlet (Rukun Tetangga/RT), then multiplied by 100%. If the result is 0%, then the SDGs Desa target is achieved. The formula is as follows:

$$\%\ \text{Human trafficking in each neighbourhood unit} = \frac{\sum \text{Human trafficking cases in each Neighbourhood Units (RT)}}{\sum \text{Neighbourhood units (RT)}} \times 100\%$$

21.3.8 Poor People Receiving Litigation and Non-litigation Legal Aid

The target for poor people receiving litigation and non-litigation legal aid is set at 100%. The indicator used is the amount of aid received by family members in the last year. Efforts that can be made include providing knowledge about the legal process to poor people, as well as providing assistance during the legal process.

To measure the percentage of poor people receiving litigation and non-litigation legal aid, the number of poor people who receive litigation and non-litigation legal aid is divided by the total number of poor people who have legal problems, then multiplied by 100%. If the result is 100%, then the SDGs Desa target is achieved. The formula is as follows:

$$\%\ \text{Poor people receiving litigation and non-litigation legal aid} = \frac{\sum \text{Poor people receiving litigation and non-litigation legal aid}}{\sum \text{total number of poor people who have legal problems}} \times 100\%$$

21.3.9 Judicial Services for the Poor Through Court Sessions Outside the Court Building

The target for judicial services for the poor through court sessions outside the court building is set at 100%. The head of household/individual indicator used is the number of family members who undergo court sessions outside the court. Efforts that can be made to increase assistance to village communities who are currently having legal problems include providing legal support for the poor, as well as conducting educational seminars and free consultations related to legal aid, and providing village paralegals for independent problem-solving by the village.

To measure the percentage of judicial services for the poor through court sessions outside the court building, the number of poor people who receive judicial services through court sessions outside the court building is divided by the total number of poor people who need judicial services through court sessions outside the court building, then multiplied by 100%. If the result is 100%, then the SDGs Desa target is achieved. The formula is as follows:

$$\begin{aligned}&\%\ \text{Judicial services for the poor through court}\\&\text{sessions outside the court building}\\&=\frac{\sum\begin{array}{l}\text{Poor people who receive judicial services}\\\text{through court sessions outside the court building}\end{array}}{\sum\begin{array}{l}\text{Poor people who need judicial services}\\\text{through court sessions outside the court building}\end{array}}\times 100\%\end{aligned}$$

21.3.10 Case Fee Waiver for the Poor

The target for case fee waivers is set to apply to 100% of the poor. The head of household/individual indicators used are the number of family members who received legal aid in the last year and the number of family members who received a case fee waiver. Efforts can be made to increase case fee waivers for the poor by, for example, allocating legal aid funds for the poor in the village.

To measure the percentage of case fee waivers for the poor, the amount of case fees waived for the poor is divided by the amount of case fees for the poor, then multiplied by 100%. If the result is 100%, then the SDGs Desa target is achieved. The formula is as follows:

$$\begin{aligned}&\%\ \text{Case fee waivers for the poor}\\&=\frac{\sum \text{Case fees waived for the poor}}{\sum \text{Case fees for the poor}}\times 100\%\end{aligned}$$

21.3.11 Legal Services Post

The target is for a Legal Services Post to be available in the village. The village indicator used is the presence of a legal services post or similar. Actions that can be carried out at the legal services post include providing legal consultation, mediation, negotiation, and out-of-court assistance for village communities.

To measure the availability of a legal services post, data on the number of legal services posts is needed. If there is a legal services post in the village, then the SDGs Desa target is achieved.

21.3.12 Anti-corruption Behaviour Index Reaches Number 5

The target for the Anti-Corruption Behaviour Index (Indeks Perilaku Anti Korupsi/IPAK) is to reach number 5. The head of household/individual indicator used is the amount of additional costs in village services. Efforts that can be made include: conducting Anti-Corruption socialisation, providing sanctions for parties proven to be corrupt, and using digital technology in public services to minimise corruption.

The measurement of the Anti-Corruption Behaviour Index is based on examples provided by the Central Bureau of Statistics:

(1) The attitude of a wife who receives extra money from her husband, outside the usual salary/income received, without questioning the origin of the money: reasonable, tends to be reasonable, tends to be unreasonable, unreasonable.
(2) The attitude of a civil servant using an official vehicle for family purposes: reasonable, tends to be reasonable, tends to be unreasonable, unreasonable.
(3) The attitude of parents inviting their child in a local election/general election campaign to get more money: reasonable, tends to be reasonable, tends to be unreasonable, unreasonable.
(4) The attitude of taking money belonging to other family members without the owner's permission: reasonable, tends to be reasonable, tends to be unreasonable, unreasonable.
(5) The attitude of using other family members' belongings without the owner's permission Owner's attitude: reasonable, tends towards reasonable, tends towards unreasonable, unreasonable.
(6) Attitude of giving money/goods/facilities to the head of Neighbourhood Units (RTs)/Community Units (RWs)/village chief/urban village head when a family is holding a celebration: reasonable, tends towards reasonable, tends towards unreasonable, unreasonable.
(7) Attitude of giving money/goods/facilities to other community leaders when a family is holding a celebration: reasonable, tends towards reasonable, tends towards unreasonable, unreasonable.

(8) Attitude of giving money/goods/facilities to the head of RT/RW/village chief/urban village head on the eve of religious holidays: reasonable, tends towards reasonable, tends towards unreasonable, unreasonable.
(9) Attitude of giving money/goods/facilities to other community leaders on the eve of religious holidays: reasonable, tends towards reasonable, tends towards unreasonable, unreasonable.
(10) Attitude of guaranteeing a family member/friend to be accepted as a civil servant/private employee in order to maintain familial and friendship relations: reasonable, tends towards reasonable, tends towards unreasonable, unreasonable.
(11) Attitude of giving money/goods/facilities in the process of being accepted as a civil servant/private employee: reasonable, tends towards reasonable, tends towards unreasonable, unreasonable.
(12) Attitude of giving money/goods/facilities to officials to expedite population administration matters: reasonable, tends towards reasonable, tends towards unreasonable, unreasonable.
(13) Attitude of giving money/goods/facilities to law enforcement officers to expedite the processing of driving licenses, vehicle registration certificates, police clearance certificates, etc.: reasonable, tends towards reasonable, tends towards unreasonable, unreasonable.
(14) Attitude of traffic violators who give peace money to the police: reasonable, tends towards reasonable, tends towards unreasonable, unreasonable.
(15) Attitude of marriage registrar officials asking for additional money for transport to the wedding ceremony venue: reasonable, tends towards reasonable, tends towards unreasonable, unreasonable.
(16) Attitude of a teacher receiving assurance (quota) that their child is accepted at the school where they teach: reasonable, tends towards reasonable, tends towards unreasonable, unreasonable.
(17) Attitude of the school (teacher/principal/school committee) asking for money/goods/facilities from parents during grade promotion/report card distribution: reasonable, tends towards reasonable, tends towards unreasonable, unreasonable.
(18) Attitude of giving money/goods/facilities to the school so that their child can be accepted at that school: reasonable, tends towards reasonable, tends towards unreasonable, unreasonable.
(19) Attitude of village/city/election participants distributing money/goods/facilities to potential voters: reasonable, tends towards reasonable, tends towards unreasonable, unreasonable.
(20) Accepting the distribution of money/goods/facilities during village/city/elections: reasonable, tends towards reasonable, tends towards unreasonable, unreasonable.
(21) Community members who have dealt with public services and have spent money/goods/facilities exceeding the provisions in the last 12 months: yes, by themselves, yes through an intermediary, no.

(22) Business operators who have dealt with public services and have spent money/goods/facilities beyond the stipulated amount: yes, by themselves, yes, through an intermediary, no.
(23) Members of the public who have dealt with public services and have never spent money/goods/facilities beyond the stipulated amount for reasons: the cost is in accordance with the regulations, refusing bribery, fear of breaking the law, not having extra money, no benefits.
(24) Officials who receive money/goods/facilities beyond the stipulated amount according to gender: male, female, don't know.
(25) Members of the public who have dealt with public services and have spent money/goods/facilities beyond the stipulated amount according to the time of payment: before the service; after the service; before, during, and after the service; at the time of service; before and after service.
(26) Members of the public who have dealt with public services and have spent money/goods/facilities beyond the stipulated amount according to the form of payment: money, food, other goods, facilities.
(27) Members of the public who have dealt with public services and have spent money/goods/facilities beyond the stipulated amount for reasons: to speed up the process, to get better service, to avoid fines, to get information, as a thank you, to maintain good relations, others.
(28) Members of the public who have dealt with public services and have spent money/goods/facilities beyond the stipulated amount according to the source: requested by the officer, requested by a third party, a common thing, no one asked.
(29) Members of the public who have dealt with public services and have spent money/goods/facilities beyond the stipulated amount according to their response: do not mind, mind.
(30) Members of the public who reported/complained because they spent money/goods/facilities beyond the stipulated amount when dealing with public services: yes, no.
(31) Members of the public who reported/complained because they spent money/goods/facilities beyond the stipulated amount when dealing with public services: pointless, will not be followed up; don't know how and to whom to report; the reporting process is complicated and long; afraid of being complicated in the next service; not reported back; others.
(32) Members of the public who reported/complained because they spent money/goods/facilities beyond the stipulated amount when dealing with public services according to the place of reporting/complaining: police, complaint service, social media, others.
(33) Members of the public who reported/complained because they spent money/goods/facilities beyond the stipulated amount when dealing with public services according to the reason for reporting/complaining: angry/disappointed with the service condition, feeling disadvantaged, wanting to help improve the situation.

(34) The experience of whether the public has been offered money/goods/facilities to choose a certain candidate in the village elections, regional elections, or the last general election: yes, no, do not remember.
(35) The response of the public who have been offered money/goods/facilities to choose a certain candidate in the village elections, regional elections, or the last general election: accept, accept reluctantly, refuse politely, refuse firmly.
(36) Whether the public has ever been offered by someone to get a household member accepted as a civil servant/private employee by having to pay a certain amount of money: yes, no, don't remember.
(37) The public's response to being offered by someone to get a household member accepted as a civil servant/private employee by having to pay a certain amount of money: accept, accept reluctantly, reject politely, reject firmly.
(38) Whether the public has ever been offered help/guarantee of acceptance by a relative/friend for a household member to become a civil servant/private employee: yes, no, don't remember.
(39) The public's response to being offered help/guarantee of acceptance by a relative/friend for a household member to become a civil servant/private employee: accept, accept reluctantly, reject politely, reject firmly.
(40) Whether the public has ever been offered help/guarantee of acceptance by a relative/friend for a household member to pass school admission selection: yes, no, don't remember.
(41) The public's response to being offered help/guarantee of acceptance by a relative/friend for a household member to pass school admission selection: accept, accept reluctantly, reject politely, reject firmly.
(42) Whether the public has ever been offered to pay a peace money when ticketed by traffic police officers: yes, no, don't remember.
(43) The public's response to being offered to pay peace money when ticketed by traffic police officers: accept, accept reluctantly, reject politely, reject firmly.

21.3.13 Financial Report Accepted by Village Deliberation Forum

In achieving the target of the last financial report being accepted in the Village Deliberation Forum, the village indicator used is the existence of a Village Deliberation Forum report on the realisation of the village budget and the acceptance of the village financial report in the Village Deliberation Forum.

21.3.14 Financial Report can be Accessed by the Public

The target for access to the financial report is that it is expected to be accessible to the public. The head of household/individual indicator used is the ease of access to village financial documents and its median (online or offline). Efforts that can be made so that the financial report can be accessed by the public include creating public publication facilities related to the use of village finances and assets that are open to the public, creating a digital data-based village financial and asset administration system, conducting financial management training for villages, and conducting village cadre training for financial management assistance.

21.3.15 Last Head of Village's Accountability Report Accepted in Village Consultation

In achieving the target of the last head of village's accountability report being accepted in the Village Deliberation Forum, the village indicator used is the head of village's accountability report accepted by the Village Deliberation Forum.

21.3.16 Procurement Process of Goods and Services for All Development Projects Open to the Public

The goal of the procurement process for goods and services for all development projects is set that they should be open to the public, translated into indicators in the village in the form of the number of procurement of goods and services and the number of open publications of procurement of goods and services in the village. Efforts that can be made so that the procurement of goods and services for all development projects is open to the public, for example by creating public publication facilities related to the procurement of goods and services for all development projects.

21.3.17 Village Government Organisational Structure and Governance in accordance with Applicable Legislation

The target of the Organisational Structure and Governance (Struktur Organisasi dan Tata Kerja/SOTK) of the village government is that it must comply with applicable legislation. The village indicators used are the conformity of the Organisational Structure and Governance of the village government with legislative rules or the number of village apparatus based on position.

The target level of citizen satisfaction with village government services is that it must be high. The head of household/individual indicators used are the level of satisfaction with village services over the past year (good, fair, poor). Efforts that can be made to improve public services for village communities are to establish integrated service posts in the village.

21.3.18 Level of Citizen Satisfaction with Village Government Services

The level of citizen satisfaction can be measured with indicators of good (valued at 3), fair (valued at 2), and poor (valued at 1). Measuring the level of citizen satisfaction, can be done by adding up the values given by village residents, divided by the number of village residents who give satisfaction level values, divided by 3, then multiplied by 100%. If the result is more than 100%, then the SDGs Desa is achieved. This measurement uses the following formula:

$$\begin{aligned}&\%\ \text{Level of citizen satisfaction with Village Government Service}\\&=\frac{1}{3}\frac{\sum \text{Value given by village residents}}{\sum \text{Village residents who give satisfaction level values}}\times 100\%\end{aligned}$$

21.3.19 Women in the Village Consultative Body

The target for the percentage of women in the Village Consultative Body (Badan Permussyawaratan Desa/BPD) is set at a minimum of 30%. The village indicators used are the number of women who attend the village deliberation forum and the number of women who are members of the Village Consultative Body (BPD). Efforts that can be made to increase the involvement of women as Village Consultative Body (BPD) members include providing space for women's participation in Village Consultative Body (BPD), as well as organising women's leadership training.

Measuring the percentage can be done by adding up the number of women in the Village Consultative Body (BPD), dividing by the number of Village Consultative Body (BPD) members, then multiplying by 100%. If the result is more than 30%, then the SDGs Desa is achieved. This measurement uses the following formula:

$$\begin{aligned}&\%\ \text{Women in the Village Consultative Body}\\&=\frac{\sum \text{Women member in Village Consultative Body}}{\sum \text{Members of the Village Consultative Body}}\times 100\%\end{aligned}$$

21.3.20 *Women in Village Government Reach a Minimum of 30%*

The target for the percentage of women in village government is set at a minimum of 30%. The village indicators used are the gender of the village head and the number of women village officials along with their positions. Efforts that can be made to increase the involvement of women as members of the village government include providing space for women's participation in village governance and conducting women's leadership training.

Measuring the percentage of women in village government can be done by adding up the women who are part of the village government, dividing by the number of members of the village government, then multiplying by 100%. If the result is more than 30%, then the SDGs Desa is achieved. This measurement uses the following formula:

$$\begin{aligned}&\%\ \text{Women in village government}\\&=\frac{\sum \text{Women who are part of the village government}}{\sum \text{Members of the village government}}\times 100\%\end{aligned}$$

21.3.21 *Democracy Institution Index*

The target for the Democracy Institution Index is set to reach 100. The head of household/individual indicator used is the satisfaction level with village institutions, such as Village Consultative Body (BPD) in the last year (good, fair, poor). Efforts that can be made to increase the number of democratic activities in the village include, for example, holding open village deliberations for the public related to the August 17th commemoration agenda and religious activities.

To measure the Democracy Institution Index, weighing data based on Analytical Hierarchy Process (AHP) from democratic institutions (P) multiplied by the democracy institution index ($I(A)$) is needed. The index result is a scale of 1–100, that is, <60: poor, 60–80: fair, and >80: good. If the result is 100, it can be said that the SDGs Desa target has been achieved. The formula used is as follows:

$$\text{Democracy Institution Index} = P \times I(A)$$

21.3.22 Civil Freedom Index

The Civil Freedom Index aims to reach 100. The head of household/individual indicators used are the level of village openness to citizen input over the past year (good, fair, poor). Efforts that can be made so that the community is free to express opinions include creating a suggestion box/input for the general public, holding open village deliberation meetings for all residents, and creating integrated village complaint service posts.

To measure the Civil Freedom Index, weighting data is needed based on the Analytical Hierarchy Process (AHP) of the civil freedom aspect (P) multiplied by the civil freedom index ($I(A)$). The result of the index is a scale of 1–100, namely <60: poor, 60–80-average, and >80: good. If the result is 100, it can be said that the SDGs Desa target has been achieved. The formula used is as follows:

$$\text{Civil Freedom Index} = P \times I(A)$$

21.3.23 Political Rights Index

The target of the Political Rights Index is to reach 100. The head of household/individual indicators used are the level of village protection of citizens' rights over the past year (good, fair, poor). Efforts that can be made to maintain the political rights of village residents include conducting campaigns/publications related to calls to participate in the election of village heads, as well as re-registering residents who are eligible to vote for the village head.

To measure the Political Rights Index, weighting data is needed based on AHP from political rights (P) multiplied by the political rights index ($I(A)$). The result of the index is a scale of 1–100, that is, <60: poor, 60–80-average, and >80: good. If the result is 100, it can be said that the SDGs Desa target has been achieved. The formula used is as follows:

$$\text{Political Rights Index} = P \times I(A)$$

21.3.24 Coverage of Birth Certificate Ownership for the Bottom 40% of the Income Population

To measure the percentage of birth certificate ownership coverage for the bottom 40% of the income population, the number of the bottom 40% of the income population who have a birth certificate is divided by the number of the bottom 40% of the income population, multiplied by 100%. If the result is 100%, the SDGs Desa target has been achieved. The formula is as follows:

$$\text{\% Coverage of Birth Certificate Ownership for the Bottom 40\% Income Population} = \frac{\sum \begin{array}{l}\text{the bottom 40\% income population}\\ \text{who have a birth certificate}\end{array}}{\sum \text{the bottom 40\% income population}} \times 100\%$$

21.3.25 *Children with Birth Certificates*

The target for birth certificate ownership coverage for the bottom 40% of the income population is set at 100%. The indicators used are the number of family members, the number of family members who have birth certificates, and the number of birth certificate processing fee aids. Efforts that can be made to encourage all communities to have birth certificates include conducting socialisation about birth certificates, conducting data collection of birth certificate ownership for the population, and providing birth certificate creation fee assistance for the less fortunate.

To measure the percentage of children with birth certificates, the number of children who have birth certificates is divided by the number of children in the village, multiplied by 100%. If the result is 100%, then the SDGs Desa target is achieved. The formula is as follows:

$$\text{\% Children with birth certificates} = \frac{\sum \text{Children who have birth certificates}}{\sum \text{Children in the village}} \times 100\%$$

21.3.26 *Handling of Complaints of Violations due to Race, Ethnicity, and Religion*

The target for handling complaints of violations due to race, ethnicity, and religion is set at 100%. The village indicators used are the number of customer complaints due to race, ethnicity, and religion as well as the number of handling of complaints of violations due to race, ethnicity, and religion. Efforts that can be made include establishing a complaint service post and mediation for handling human rights violation conflicts, as well as providing financial assistance for legal advocacy related to human rights violations.

To measure the percentage of handling of violations due to race, ethnicity, and religion, the number of complaints due to race, ethnicity, and religion that are handled is divided by the total number of complaints due to race, ethnicity, and religion, then multiplied by 100%. If the result is 100%, then the SDGs Desa target is achieved. The formula is as follows:

$$\%\ \text{Handling of violations due to race, ethnicity, and religion} = \frac{\sum \begin{array}{c}\text{Complaints due to race, ethnicity,}\\ \text{and religion that are handled}\end{array}}{\sum \text{Complaints due to race, ethnicity, and religion}} \times 100\%$$

21.3.27 Development Planning Documents and Financial Report Documents can be Accessed by the Public

The target for development planning documents and financial report documents is that they must be publicly accessible up to 100%. The indicator used is the ease of access for households/individuals. Village development and financial planning documents (good, average, poor). Efforts that can be made include publicising the use of the village budget and all ongoing activities in the village, creating public facilities related to the use of village finances and assets that are open to the public, and creating a village information system based on digital data.

This aspect is measured using the ease of access to village development and financial planning documents, namely good (worth 3), average (worth 2), and poor (worth 1). Measuring the aspect can be done by adding up the values given by the villagers, dividing by the number of villagers who gave the value, dividing by 3, then multiplying by 100%. If the result is more than 100%, then the SDGs Desa are achieved. This measurement uses the following formula:

$$\begin{aligned}&\%\ \text{Development planning documents and financial report}\\&\text{documents can be accessed by the public}\\&\quad = \frac{1}{3}\frac{\sum \text{Value given by the villagers}}{\sum \text{Villagers who give value}} \times 100\%\end{aligned}$$

21.3.28 Summary of Development Planning Documents and Financial Report Documents Announced to the Public

The target of the summary of development planning documents and financial report documents should be announced to the public in an open space. The village indicator used is the announcement in an open space of the summary of the village's development planning and financial documents.

21.3.29 Resolution of Complaints Related to Public Information

The target for percentage of resolution of complaints related to public information is set at 100%. The village indicator used is the number of village public information dispute complaints in the last year and the number of resolutions of village public information dispute complaints in the last year. Efforts that can be made to reach this target include: building integrated complaint service posts and mediation for conflict management.

To measure the percentage, the number of complaints related to public information that are resolved is divided by the total number of complaints related to public information, and then multiplied by 100%. If the result is 100%, then the SDGs Desa target is achieved. The formula is as follows:

$$\%\ \text{Resolution of Complaints Related to Public Information}$$
$$= \frac{\sum \begin{array}{l}\text{Complaints related to public}\\ \text{information that are resolved}\end{array}}{\sum \text{Complaints related to public information}} \times 100\%$$

21.3.30 Provision of Information on the Same Day

The indicator of households/individuals used to show the target of information provision by the village government on the same day is the number of days the village government provided village information requests over the past year and the number of days the village government provided village information requests the previous year. Efforts that can be made to speed up the provision of information include creating a financial and asset administration system based on digital data, as well as providing training in managing the village information system digitally.

Table 21.1 Operational definitions for village of peace and justice

	Key word	Operational definition
1	Village assets	Property owned by the village that originates from the wealth of the village, purchased or acquired at the expense of the village revenue and expenditure budget and other legitimate rights acquired
2	Political rights index	The extent to which citizens have the right to participate in government activities
3	Democracy institution index	A composite measure that evaluates the capacity, performance, and role of various state institutions in supporting and implementing democratic governance. It examines whether these institutions operate effectively, impartially, and in accordance with democratic norms and the rule of law.
4	Anti-corruption behaviour index (IPAK)	A measure of society's increasingly anti-corruption behaviour
5	Domestic violence (KDRT)	All behaviours of threat, harassment, and violence, whether physical, psychological, or sexual, between two people in a personal relationship or towards other family members
6	Conflict	Disputes between community members
7	Criminality	Any act or action committed by an individual or several people at once that violates criminal law
8	Information and documentation management officer (PPID)	The manager and deliverer of documents owned by a public body
9	Village information system	An application used to manage various data in the village

21.4 Operational Definitions for Village of Peace and Justice

Key words related to the Village of Peace and Justice are presented in Table 21.1.

References

UN. n.d.(a). *Goal 12: Ensure sustainable consumption and production patterns* [internet].

UN. n.d.(b). *Goal 15* [internet]. https://sdgs.un.org/goals/goal15

UN. n.d.(c). *Goal 16: Promote just, peaceful and inclusive societies* [internet]. https://www.un.org/sustainabledevelopment/peace-justice/

UN. n.d.(d). *Life below water: Why it matters* [internet]. https://www.un.org/sustainabledevelopment/wp-content/uploads/2019/07/14_Why-It-Matters-2020.pdf

Chapter 22
SDGs Desa 17: Partnerships for Village Development

22.1 Concept of Partnerships for Village Development

Partnership, according to the Great Dictionary of the Indonesian Language, means a relationship, collaboration, and similar, as a partner. Partnerships for Village Development becomes a means of implementation and revitalisation of village partnerships to achieve all sustainable development goals. Village development will not be maximally successful without the involvement of related parties, ranging from community leaders, village youth activists, village economic women activists, universities, the business world, supra village, of course also village apparatus and the Village Consultative Body (Badan Permusyawaratan Desa/BPD).

The world has never been more interconnected than it is now. Improving access to technology and knowledge is a crucial way to share ideas and drive innovation. Currently, in the post-Covid-19 pandemic era, strong cooperation is needed, more than ever, to ensure villages have the means to recover from the pandemic, rebuild better, and achieve SDGs Desa. This goal targets the strengthening of cooperation to achieve growth and sustainable development. To measure the achievement of this goal, several performance indicators are used, including the existence and form of village cooperation with third parties, the availability of internet networks in villages, village statistics, and commodities and export activities by the village.

22.2 Partnerships for Village Development Targets

To achieve the Partnerships for Village Development goal, the following 17 targets have been set to be achieved by 2030:

(1) The ratio of tax revenue to Village GDP is above 12% per year

(1) There is cooperation between villages
(2) There is cooperation between the village and the private sector

A. H. Iskandar, *The Methods of Measuring the National Sustainable Development Goals in Indonesia*, Engaging Indonesia, https://doi.org/10.1007/978-981-96-0927-7_22

(3) There is cooperation with NGOs
(4) There is cooperation with international institutions
(5) Availability of internet networks in the village
(6) Fast category internet speed
(7) There is Wi-Fi in the village
(8) Internet speed using fast category Wi-Fi
(9) 100% internet usage with mobile phones
(10) Internet speed using mobile phones in the fast category
(11) Information on the economic conditions of the village can be accessed by the public
(12) Village statistical data is available every year
(13) There are special officers handling village statistics
(14) There is a village statistics application
(15) Annual SDGs Desa data is available
(16) Use of SDGs Desa data for Village Government Workplan (Rencana Kerja Pemerintah/RKPDes) and Village Revenue and Expenditure Budget (Anggaran Pendapatan dan Belanja Desa/APBDes).

22.3 Measurement of Partnerships for Village Development

22.3.1 Ratio of Tax Revenue to Village GDP per Year

The target ratio of tax revenue to village GDP is set at above 12% per year. The village indicators used are the amount of land and building tax (Pajak Bumi dan Bangunan/PBB) obligations this year and last year, and the amount of land and building tax payments this year and last year. Efforts that can be made to increase the amount of business tax and village community land and building tax include building public consultation spaces for taxes, carrying out data collection, creating drone-based land and building tax maps, and conducting tax payment campaigns.

To measure the ratio of tax revenue to village GDP, the required data is the result of local tax and regional retribution (D.P601.3.) divided by village funds (D.P601.1), PADes (D.P601.2), Village Fund Allocation (D.P601.4), financial assistance from the Provincial APBD (D.P601.5), financial assistance from the District/City APBD (D.P601.6), grants (D.P601.7), and others (D.P601.8) from the last year, and then multiplied by 100%. If the result is above 12%, then the SDGs Desa target is achieved. The formula is as follows:

$$\%\ \textit{The result of local tax and regional retribution to village GDP}$$
$$= \frac{\mathbf{D.P}601.3}{\mathbf{D.P}601.1 + \mathbf{D.P}601.2 + \ldots + \mathbf{D.P}601.8} 100\%$$

22.3.2 Inter-village Cooperation, with Third Parties, and International Cooperation

The next target is the existence of inter-village cooperation, with private sector, NGOs, and international institutions. The village indicators used are the number and type of inter-village cooperation, the number and type of village cooperation with the private sector, the number and type of village cooperation with NGOs, as well as the number and type of village cooperation with international institutions in this year and the previous year. Efforts that can be made to develop partnerships with multiple parties (universities, private sector, international institutions, ministries/agencies, NGOs, inter-regional) include: carrying out Thematic Community Service Programme development, in cooperation with universities, to develop the village's potential, or solve problems in the village; working with *offtaker* in the development and marketing of flagship products (for example, for the development of copra exports); working together with the private sector in supporting the education sector (e-learning, learning methods) and health (e-health, vaccine delivery, medicines, and blood using drone); implementing cooperation with the online store marketplace; and conducting collaboration with startup as aggregator (collector of village flagship products/commodities).

The existence of village cooperation with international institutions is indicated by the number of instances of village cooperation with international institutions this year and the number of instances of village cooperation with international institutions last year. Efforts that can be made to promote the village and offer cooperation for the development of village potential are to socialise the village's potential digitally and promote the village's flagship products.

To measure the existence of village cooperation with other villages, third parties, and international institutions, village cooperation data (D.P705) is needed. If the village has participated in cooperative activities, then the SDGs Desa target is achieved. The formula is as follows:

$$\begin{aligned} &\textit{The existence of village cooperation with other villages,}\\ &\textit{third parties, and international institutions}\\ &= (\boldsymbol{D.P}705.1 \neq 0 \cup \ldots \cup \boldsymbol{D.P}705.7 \neq 0) \end{aligned}$$

22.3.3 Internet Network in the Village

The target for the village internet network is for it to be available by 2030. The village indicator used is the presence of a village internet network. Efforts that can be made to build communication and internet networks for villages without signal and internet include building a village internet network and its supporting facilities, installing Wi-Fi in public village spaces, cooperating for the construction of telephone signal networks in villages without signal, paying for the maintenance of internet in the village, and providing training in the management of village telephone and internet networks.

To achieve the target of fast internet speed, the village indicator used is the internet signal in the village (very strong signal, strong signal, weak signal, and no signal).

To achieve the target of Wi-Fi presence in the village, the indicator from the family/individual questionnaire used is the use of internet with Wi-Fi (in a month) over the past year.

To achieve the target of internet speed using Wi-Fi, the family/individual questionnaire used is the internet signal using Wi-Fi (very strong signal, strong signal, weak signal, and no signal).

To achieve the target of 100% internet usage with mobile phones, the family/individual questionnaire used is the use of internet with mobile phones (in a month) over the past year.

The target of internet speed using mobile phones, indicated from the family/individual questionnaire, is that the internet signal using mobile phones is used (very strong signal, strong signal, weak signal, and no signal).

To measure the availability of a fixed internet network (Wi-Fi) and mobile (mobile phone) with high speed, data on village internet facilities (D.P701) and mobile signal (R.P607) are needed. If a high-speed fixed (Wi-Fi) and mobile (mobile phone) internet network is available, then the SDGs Desa target is achieved. The formula is as follows:

$$\begin{aligned}&\textit{The availability of a fixed internet network } (\textit{wi} - \textit{fi}) \textit{ and}\\&\textit{mobile } (\textit{mobile phone}) \textit{ with high speed}\\&= (\mathbf{D}.\mathbf{P}701.2 = 1 \cap \mathbf{R}.\mathbf{P}607 = 1|2)\end{aligned}$$

22.3.4 Commodity Exports from the Village

To check the increase in commodity exports from the village, the indicators used are the type, quantity, and value of commodities exported in the last year and the type, quantity, and value of commodities exported the previous year. Efforts that can be made to build cooperation for exporting the village's flagship commodities are to market them *online* in international markets (for example through Alibaba), join export network communities, build supporting facilities for cooperation (for example, building drying houses to support copra exports in cooperation with *offtaker* incaom.id), and to work in cooperation with the development of flagship commodities with *offtaker* (for example, lab test costs and research and development).

To measure the village commodities that are exported, data on the income of family members who have exported in the last year and the previous year (I/P204.export) are needed. If the result increases, then the SDGs Desa target is achieved. The formula is as follows:

$$\textit{The number of resident who export their commodities} = \sum (\mathbf{I.P}204.\mathbf{export} = 1|2)$$

22.3.5 Existence of Village Information System

To achieve the target of village economic condition information, the village indicator used is the availability of village economic information. Efforts that can be made are to build a village website as a centre for village data and information and to conduct village website management training.

To measure the existence of the village information system and the updated village financial system, the existence of the village information system and the village financial system data are needed. If both of these systems exist, then the SDGs Desa target is achieved. The formula is as follows:

$$\textit{The updated village information system} = \textit{if}\,(\mathbf{D}.\mathbf{P}702.1 = 1)$$

$$\textit{The updated village financial system} = \textit{if}\,(\mathbf{D}.\mathbf{P}702.2 = 1)$$

22.3.6 Existence of Village Statistical Data, Staff and Applications

Village statistical data is targeted to be available every year. The village indicator used is the availability of village data. Efforts that can be made are to build a website for the village as a data and information centre, carry out village statistical data collection (village potential, population, and so on), and finance the creation of an integrated village website.

A special officer handling village statistic is targeted to be available. The village indicator used is the number of special officers handling village statistics. Efforts that can be made are to train officers to be able to manage village statistical data through appropriate training courses.

A village statistical application is targeted to be available. The village indicator used is the availability of a village statistical application. Efforts that can be made are to build a village website as a data and information centre for the village and as a service centre for village residents and promote the use of the village website as a means of complaint service or other population services.

22.3.7 Existence of SDGs Desa Data

Annual SDGs Desa Data is targeted to be available. The village indicator used is the availability of annual SDGs Desa Data. Efforts that can be made are to build a village website as a data and information centre, carry out village statistical data collection (village potential, population, and so on), and conduct training in managing village statistical data.

22.3.8 RKP Desa and APB Desa Based on SDGs Desa Data

SDGs Desa Data is targeted to be used for RKP Desa and APB Desa. The village indicator is the use of SDGs Desa Data for RKP Desa and APB Desa. Efforts that can be made are to build a website for the village as a centre for village data and information, to collect village statistics (village potential, population, and so on), as well as conducting training in village statistical data management.

22.4 Operational Definitions for Partnerships for Village Development

Key words related to Partnerships for Village Development are presented in Table 22.1.

Table 22.1 Operational definitions for partnerships for village development

Key word	Operational definition
Export	**The activity of selling goods abroad**
Mobile phone	An electronic telecommunications device that has the same basic capabilities as a conventional fixed-line telephone but can be carried anywhere
International institution	An organisation created by members of the international community
Tax	A source of state financial income by collecting funds or money from taxpayers for the state treasury to finance national development or other state expenditures
Land tax map	A map created to provide spatial information including, street names, object location, and attribute information including tax object number, taxpayer name, tax object address, taxpayer address (NJOP), Land Value Zone Code (ZNT), land area, building area, Land Tax Object Sale Value, Building Tax Object Sale Value (NJOP) and Total Tax Object Sale Value (NJOP)
Startup	A nascent business that is expected to grow and dominate a market niche quickly and subsequently become a large company
Statistics	Knowledge related to the methods of data collection, data processing, analysis, and also drawing conclusions based on a set of data and the analysis carried out
Wi-fi	Technology that uses electronic equipment to exchange data wirelessly (using radio waves) through a computer network, including high-speed internet connections

Chapter 23
SDGs Desa 18: Dynamic Village Institutions and Adaptive Village Culture

23.1 Concept of Dynamic Village Institutions and Adaptive Village Culture

Village institutions are a vital component to drive sustainable development. Therefore, the existence of village institutions must be strengthened so that they can support the achievement of sustainable development goals.

SDGs Desa 18 strives to maintain local wisdom, as well as carry out revitalization and mobilize all elements of institutions at the village level, because the involvement of all village elements, strong and functional institutions in the village in community life, will become a support for the diversity of life in a dynamic village, as well as a driver for the achievement of SDGs Desa.

To achieve the goals of SDGs Desa by 2030, several indicators are used, including the sustainability of mutual aid and cooperation, participation of religious leaders in village development activities, protection by villagers of the vulnerable, the poor and orphans, preservation of village culture, and problem-solving based on a cultural approach.

23.2 Targets for Village Dynamic Institutions and Adaptive Village Culture

The following 15 targets for SDGs Desa goal 18 have been set to be achieved by 2030:

(1) Social piety
(2) Role of religious leaders in the Village Deliberation Forum and implementation of village development
(3) Villagers concerned in caring for orphans and the poor

A. H. Iskandar, *The Methods of Measuring the National Sustainable Development Goals in Indonesia*, Engaging Indonesia, https://doi.org/10.1007/978-981-96-0927-7_23

(4) Organisational structure and governance of village government in accordance with applicable laws and regulations
(5) Implementation of Village Deliberation Forum at least four times a year
(6) Availability of RPJM Desa
(7) Availability of village boundary map determined by regent/mayor
(8) Availability of Village Government Work Plan (Rencana Kerja Pemerintah Desa/RKPDes) for the current year's implementation
(9) Availability of Village Revenue and Expenditure Budget (Anggaran Pendapatan dan Belanja Desa/APBDes) for the current year's implementation
(10) Accreditation level of Village-owned Enterprises/Joint Village-owned Enterprises (Badan Usaha Milik Desa/Bersama –BUMDes/ma) at least equivalent to 'good'
(11) Culture preserved reaches 100%
(12) Existence and participation of traditional institutions in village development
(13) Resolution of social problems based on local culture reaches more than 50%
(14) Value of village assets increases every year
(15) The percentage of community institutions participating in the Village Deliberation Forum reaches more than 30%

23.3 Measurement of Village Dynamic Institutions and Adaptive Village Culture

23.3.1 Social Piety

Social piety is a mutual aid activity based on religious teachings. The indicators used are the number of mutual aid activities with neighbours in the last six months and the reasons for helping neighbours (religion, living close, principle, politics, and so on). Efforts that can be made to improve communication and interaction in rural communities include building public spaces in the village (such as sports fields, community halls, art studios, and creative economic spaces).

To measure the awareness of residents over 15 years old who participate in mutual aid activities (death, illness, and accidents), the number of residents who help residents who have experienced death (I.P507), the number of residents who help sick residents (I.P508), and the number of residents who help residents who have had accidents (I.P509) and are aged over 15 years, is divided by the number of residents aged 15 years then multiplied by 100%. If the result is close to 100%, then the SDGs Desa are achieved. The formula is as follows.

$$\% \textit{ the awareness of residents over } 15 \textit{ years old who participate in mutual aid activities } (\textit{death, illness, and accidents})$$

$$= \frac{\sum\left[(\mathbf{I.P507} \neq 0 \cup \mathbf{I.P508} \neq 0 \cup \mathbf{I.P508} \neq 0) \cap \mathbf{I.Age} \geq 15 \text{ years}\right]}{\sum(\mathbf{I.Age} \geq 15 \text{ years})} \times 100\%$$

23.3.2 Role of Religious Leaders in the Village Deliberation Forum and Implementation of Village Development

The next target is the role of religious leaders in village meetings and the implementation of village development. The village indicator used is the number of religious leaders present at the village meetings. The Neighbourhood Units/hamlets (Rukun Tetangga/RT) indicator used is the number of religious leaders participating in village development activities. Efforts that can be made are to introduce and disseminate information about rules, planning, implementation, and monitoring of village development to increase community participation through: the development of village development information openness, socialization meetings (religious forums, customs, and so on), print media, audio-visual, and others.

23.3.3 Community Concern for Orphans and the Poor

Village residents are also expected to care for the maintenance of orphans and the poor, measured by the number of charity/maintenance activities for orphans and the poor. The hamlets (Rukun Tetangga/RT) indicator used is the existence of institutions that provide charity/maintenance for orphans and the poor. The head of household/individual indicator used is activities that provide charity/maintenance for orphans and the poor. Efforts that can be made to improve the quality of village development planning to accelerate poverty alleviation include compiling data and evaluating integrated poverty; compiling planning and budgeting that favours the poor, children and marginal groups; and fulfilling access to basic services, for example, by building/improving houses for the poor, providing school companion assistance for the poor, providing educational equipment assistance for the poor before children are accepted in educational units, and providing educational cost assistance (at least at the secondary education level) for children from vulnerable families . Another effort is to increase the economic income of poor families through the formation and training of community empowerment cadres and the provision of supporting economic infrastructure, such as training centres, production centres, and marketing for the poor.

To measure the percentage of indigence certificates issued against the number of poor people, the number of indigence certificates (D.P703) is divided by the number of poor residents then multiplied by 100% is needed. If the result is close to 100%, then the SDGs Desa target is achieved. The formula is as follows:

$$\begin{aligned}&\%\ \text{Indigence certificates issued against the number of poor people}\\&= \frac{\text{D.P703}}{\text{The number of poor residents}} \times 100\%\end{aligned}$$

23.3.4 Village Government SOTK in Accordance with Legislation

The Organisational Structure and Governance of the village government must comply with the applicable laws and regulations. The village indicator used is the conformity of the village government's Organizational Structure and Governance with laws and regulations or the number of village officials based on position. The head of household/individual indicator used is the level of satisfaction with village services in the past year (good, fair, poor). Efforts that can be made to improve the quality of village services include: compiling a profile of village officials and their job descriptions, compiling Standard Operating Procedures for village services, and conducting surveys and evaluations of village services regularly.

To measure the percentage of completeness of village official data, 94 data entries of village officials (D.P301–D.P394) are needed. If each data point is available, then it is counted as 1 (one). The total data that is available is then divided by 94 and multiplied by 100%. If the result is more than 45%, then it is considered good and complete. The formula is as follows.

$$\begin{aligned}&\%\ \mathit{Completeness\ of\ village\ official\ data}\\&=\frac{\mathbf{D}.\mathbf{P}301\neq 0+\mathbf{D}.\mathbf{P}301\neq 0+\ldots+\mathbf{D}.\mathbf{P}394\neq 0}{94}\times 100\%\end{aligned}$$

23.3.5 Implementation of the Village Deliberation Forum at Least Four Times a Year

The Village Deliberation Forum, known as Musyawarah Desa (Musdes), should be held at least four times a year. The village indicator used is the number of Village Deliberation Forums in a year. Efforts that can be made to develop participatory village development planning are to develop the openness of village development information through socialisation meetings, print media, audio-visual, and other means, as well as documenting Village Deliberation Forum activities (notes, participants, photos, and others).

To measure the percentage for the implementation of at least four Village Deliberation Forums annually, data on the completion of the first to fourth Village Deliberation Forums (D.P401.1-D.P401.4) during the year is needed. Each Village Deliberation Forum that takes place is counted as 1 (one). The number of Village Deliberation Forums that took place is divided by four then multiplied by 100%. If the result is 100% or more than 100%, then the SDGs Desa target is achieved. The formula is as follows:

$$\% \, Musdes \, implementation \, at \, lesat \, four \, time \, a \, year = \frac{\mathbf{D.P}401.1 \neq 0 + \mathbf{D.P}401.2 \neq 0 + \mathbf{D.P}401.3 + \mathbf{D.P}401.4}{4} \times 100\%$$

23.3.6 *Availability of Village Budget and Development Work Plan Documents*

The RPJM Desa must exist. The village indicator used is the existence of the Village Medium Term Development Plan (Rencana Pembangunan Jangka Menengah Desa/RPJM Desa). Efforts that can be made to determine and establish strategic issues and village development priorities are creating a timeline and activity plan for the preparation of the Village Medium Term Development Plan and preparing data and information for the needs of preparing the Village Medium Term Development Plan.

To measure the availability of the Village Medium Term Development Plan (RPJMDes), Village Government Work Plan (RKPDes), and Village Revenue and Expenditure Budget (APBDes), the documents RPJM Desa data (D.P504), RKP Desa (D.P505), and Village Funds (D.P601.1) are needed. Each available document is counted as 1 (one) then divided by three and multiplied by 100%. If the result is 100%, or all documents are available, then the SDGs Desa target is achieved. The formula is as follows:

$$\% \, The \, availability \, of \, RPJMDes, \, RKPDes, \, and \, APBDes \, documents = \frac{\mathbf{D.P}504 \neq 0 + \mathbf{D.P}505 \neq 0 + \mathbf{D.P}601.1 \neq 0}{3} \times 100\%$$

23.3.7 *Existence of Village Boundary Map*

The village boundary map that has been determined by the regent/mayor must exist. The village indicator used is the existence of a village boundary map that has been determined by the regent/mayor. Efforts that can be made to create orderly administration of government, provide clarity and legal certainty to the boundaries of a village include preparing data and supporting information for the preparation of the village boundary map.

23.3.8 Availability of Village Government Work Plan for Current Year Implementation

The Village Government Work Plan (Rencana Kerja Pemerintah Desa/RKPDes) for the current year implementation must exist. The village indicator used is the existence of RKPDes for the current year implementation. Efforts that can be made to establish a reference framework for the village head and village apparatus in preparing annual programs and activities to realize the vision and mission set in the Village Medium Term Development Plan (RPJMDes) are to create a timeline and activity plan for the preparation of RKPDes and prepare data and information for the needs of RKPDes preparation, such as indicative ceilings and data of programmes entering the village, both sourced from the regional budget and the state budget.

23.3.9 Availability of Village Revenue and Expenditure Budget for Current Year Implementation

The Village Revenue and Expenditure Budget (Anggaran Pendapatan dan Belanja Desa/APBDes) for the current year's implementation must be present. The village indicator used is the presence of APBDes for the current year's implementation. Efforts that can be made to determine the priority scale and operational development of the village to achieve the goals and targets that have been set include:creating a timeline and activity plan for the preparation of APBDes, preparing data and information for the needs of APBDes preparation, and carrying out socialisation of the current year's APBDes (offline and online documents).

23.3.10 Accreditation Level of Village-owned Enterprise at Least Equivalent Good

The Village-owned Enterprises/Joint Village-owned Enterprises (Badan Usaha Milik Desa/Bersama – BUMDes/ma) accreditation level is targeted to be at least equivalent to 'good'. The village indicator used is the BUMDes/ma accreditation level. Efforts that can be made to improve the performance and management of BUMDes/ma are to analyse and evaluate the development level of BUMDes/ma.

To measure the presence of social media in promoting BUMDes, website BUMDes data (D.P903), BUMDes' Facebook (D.P904), BUMDes' Twitter (D.P905), and BUMDes' YouTube (D.P907) are needed. Each social media has a value of 1 (one). These points are added up, divided by four, then multiplied by 100%. If the result is 100%, then the SDGs Desa target is achieved. The formula is as follows:

$$\% \ Presence\ of\ social\ media\ in\ promoting\ BUMDes\ (Web,\ Facebook,\ Twitter,\ and\ YouTube) = \frac{\mathbf{D}.\mathbf{P}903 \neq 0 + \mathbf{D}.\mathbf{P}904 \neq 0 + \mathbf{D}.\mathbf{P}905 \neq 0 + \mathbf{D}.\mathbf{P}907 \neq 0}{4} \times 100\%$$

23.3.11 Culture Preserved Reaches 100%

The culture to be preserved is targeted at 100%. The Neighbourhood Units/hamlets (Rukun Tetangga/RT) indicator used is the number of artifact locations maintained and the number of cultural activities still active in the last year. Efforts that can be made to maintain, preserve and revitalise culture are: organising art and cultural performances, building art studios, procuring and maintaining art and cultural equipment, creating village galleries/museums, providing empowerment assistance in the field of art and culture, and conducting art and culture training and courses.

To measure the number of cultural heritage sites, the number of cultural heritage sites still preserved (R.P1005) is needed. If the existing cultural heritage is still preserved, then the SDGs Desa target is achieved. The formula is as follows:

$$The\ number\ of\ cultural\ heritage\ sites = \boldsymbol{R}.\boldsymbol{P}1005.1 \neq 0 + \boldsymbol{R}.\boldsymbol{P}1005.2 \neq 0 + \boldsymbol{R}.\boldsymbol{P}1005.3 \neq 0$$

23.3.12 Presence and Participation of Traditional Institutions in Village Development

Traditional institutions must exist and participate in village development. Efforts that can be made to increase the participation of indigenous people in village development planning include: developing the openness of village development information through traditional meetings/forums, print media, audio-visual, and others; renovating traditional houses; and developing planning and budgeting for village development which is affirmative towards indigenous groups.

23.3.13 Resolution of Social Problems Based on Local Culture Achieving More Than 50%

Social problems based on local culture are targeted for achieving resolution at more than 50%. The village and Neighbourhood Units/hamlets (Rukun Tetangga/RT) indicators used are the number of social problems in the last three months and the method of resolving social problems in the last three months (e.g., according to local culture, legal, or out-of-court legal resolution). Efforts made to preserve customs and culture in maintaining peace development and diversity is to develop planning and village budgeting that is sensitive to social conflict issues and to build a community hall as a space for citizen discussion.

To measure the resolution of social problems through a cultural approach, the number of mass brawl incidents that have been resolved (R.P1101.resolution) is divided by the total number of mass brawl incidents (R.P1101) and then multiplied by 100%. If the result is above 50%, then the SDGs Desa target is achieved. The formula is as follows.

$$\% \mathit{Resolution\ of\ mass\ brawl\ incidents} = \frac{\begin{array}{c}(\boldsymbol{R.P}1101.\boldsymbol{cause}.1 \neq 0 \cap \boldsymbol{R.P}1101.\boldsymbol{resolution}.1 = 1|2) \\ \sum (\boldsymbol{R.P}1101.\boldsymbol{cause}.2 \neq 0 \cap \boldsymbol{R.P}1101.\boldsymbol{resolution}.2 = 1|2) \\ (\boldsymbol{R.P}1101.\boldsymbol{cause}.7 \neq 0 \cap \boldsymbol{R.P}1101.\boldsymbol{resolution}.7 = 1|2)\end{array}}{\begin{array}{c}\sum (\boldsymbol{R.P}1101.\boldsymbol{cause}.1 \neq 0 | R.P1101.resolution.2 \neq 0 | \ldots \\ \ldots | (R.P1101.cause.7 \neq 0)\end{array}} 100\%$$

23.3.14 The Value of Village Assets Increases Every Year

The value of village assets is targeted to increase every year. The village indicators used are the types of village assets (movable and immovable) and their value. Efforts made to secure and optimise the utilisation of village assets include inventorying potential and village assets, compiling a map of village assets, developing a digital-based village asset management administration system, and conducting village asset management training.

To measure the number of village assets, data on the number of village assets (D.P603) is needed. If the value increases from the previous year, then the SDGs Desa target is achieved. The formula is as follows:

The number of village assets = $\mathbf{D.P}603.1 + \mathbf{D.P}603.2 + \ldots + \mathbf{D.P}603.11$

23.3.15 Percentage of Village Community Institutions Participating in the Village Deliberation Forum Reaches More Than 30%

The participation of village community institutions in the Village Deliberation Forum is targeted to be more than 30%. The village indicators used are the number of

Table 23.1 Operational definitions for dynamic village institutions and adaptive village culture

Keyword	Operational definition
Religious figure	A figure who has influence in the community in religious aspects
Village asset map	A map created to provide information on village assets
Village boundary map	A map created to provide information on village boundaries

community institutions participating in the Village Deliberation Forum. The Neighbourhood Units/hamlets (Rukun Tetangga/RT) indicators used are the number of village community institutions. Efforts that can be made to increase the participation of village community institutions in planning, implementing, and supervising village development include developing openness of village development information meetings through socialisation, print media, audio-visual, and others.

To measure the percentage of village community institutions participating in the Village Deliberation Forum, the number of village community institutions (D.P800) that participate in the Village Deliberation Forum is divided by the number of community institutions and then multiplied by 100%. If the result is above 30%, then the SDGs Desa target is achieved. The formula is as follows:

$$\%\ \textit{The participation of village community insitituons in village deliberation forum}$$
$$= \frac{(\mathbf{D.P}800.1 \neq 0 + \mathbf{D.P}800.2 \neq 0 + \ldots + \mathbf{D.P}800.8 \neq 0)\mathbf{participatinginvillagedeliberationforum}}{\mathbf{D.P}800.1 \neq 0 + \mathbf{D.P}800.2 \neq 0 + \ldots + \mathbf{D.P}800.8 \neq 0} \times 100\%$$

23.4 Operational Definitions for Dynamic Village Institutions and Adaptive Village Culture

Key words related to Dynamic Village Institutions and Adaptive Village Culture are presented in Table 23.1.

Chapter 24
Field Approach

The steps to run the participatory census of SDGs Desa are as follows. First, a team for each Neighbourhood Unit/hamlets (Rukun Tetangga/RT) is formed, coordinated up to the village level. The village head can be responsible, while the team leader should be held by the village secretary. The team members can consist of village officials, including Neighbourhood Units/hamlets (Rukun Tetangga/RT) officials, data collectors who live in the same Neighbourhood Unit/hamlets (Rukun Tetangga/RT), numbering three people or odd on parties who sympathise with the village.

The village questionnaire filler is a village official assigned to collect data and information so that they can fill out the village questionnaire.

The Neighbourhood Unit/hamlets (Rukun Tetangga/RT) questionnaire filler is the /hamlets (Rukun Tetangga/RT) manager who is assigned to collect data and information so that they can fill out the Neighbourhood Unit/hamlets (Rukun Tetangga/RT) questionnaire.

The family and resident questionnaire filler is the village data volunteer assigned in each hamlets (Rukun Tetangga/RT) to interview families to fill out the family questionnaire and interview residents in order to fill out the resident questionnaire.

The team can function independently using this book, seek knowledge through social media, or set up training with village companion speakers, Central Statistics Agency officers, or students.

The village secretary prepares initial data that includes the names and addresses of families and village residents (By-Name-By-Address or BNBA), and includes the following data:

(1) Village residents who are sick, according to the type of disease; village residents who use modern family planning methods; and stunting in infants, toddlers, and children (under 15 years old). Initial data is obtained from the local Health Centre and Auxiliary Health Centre that serve the local village, as well as from local Polindes, Poskesdes, and Posyandu in the village.

A. H. Iskandar, *The Methods of Measuring the National Sustainable Development Goals in Indonesia*, Engaging Indonesia, https://doi.org/10.1007/978-981-96-0927-7_24

(2) School accreditation; the number of students and teachers at the Early Childhood Education, kindergarten, primary school, junior high school, senior high school, and equivalent levels in the local village.
(3) List of residents who participate in educational level equivalency activities in the local village, as well as a list of residents who participate in workforce training.
(4) List of residents who participate in village development and community empowerment activities.

When conducting interviews with families and residents, the following operational definitions need to be considered. The family is included in the 'card family' that is used in the family questionnaire application. A household is characterised by residents eating from one kitchen. For example, if there is a college student who is boarding, then his/her family is according to the family card, while his/her household is according to the source of breakfast, lunch, and dinner.

The enumerator must visit the family or residents more than once, if the interview is not yet complete and correct, or because the respondent is difficult to find. The enumerator must also carefully arrange an appropriate time for a return visit, choosing a time when the respondent can be found and interviewed. The enumerators must not fill in the SDGs Desa Census themselves with guesses, or estimates, or their own knowledge. All questions in the questionnaire (except where there are instructions for observation) must be asked directly to the residents. It is important not to mention only part of the questionnaire, because it can result in incomplete answers.

The Village Consultative Body (BPD), together with the village head, determine the data of the current year's SDGs Desa through village deliberation forum. The organizer of the village deliberation forum is the BPD, as a follow-up to the written notification from the Village Head that the SDGs Desa census has been completed carried out. The participants of the village deliberation forum consist of the village government, BPD, village community institutions (Lembaga Kemasyarakatan Desa/ LKD) and community elements.

At present, the processing of SDGs Desa data can be done with the help of an information system. Algorithms are built to process data according to the formula for each processed indicator.

The information obtained from the analysis of SDGs Desa data includes:

(1) By-name-by-address of poor and extremely poor residents/families, including needs according to indicators: improvement of uninhabitable houses (RTLH), electricity, drinking water, clean water, sanitation, waste services and waste management
(2) Special needs of poor, extremely poor, women, children
(3) Assistance for poor residents
(4) Food needs
(5) Types of diseases, drug use, and health facility needs
(6) Maternal and child health
(7) Length of schooling
(8) Gross Enrolment Rate, Net Enrolment Rate according to the level of education in the village

(9) Village library
(10) Basic, secondary, higher education needs
(11) Health insurance and employment needs
(12) Telematics needs
(13) Policies for women
(14) Resolution of incidents of violence against women and children
(15) Percentage of women in village government
(16) Electricity and renewable energy
(17) Unemployment rate
(18) Capital in the village
(19) Condition of land and water infrastructure
(20) Level of village advancement
(21) Security conditions in the village
(22) Environmental preservation policies
(23) Marine conservation areas
(24) Sustainable farming areas
(25) Disaster mitigation
(26) Condition of illegal fishing
(27) Condition of illegal logging
(28) Nature conservation activities
(29) Condition of conflicts in the village
(30) Condition of mutual cooperation in the village
(31) Child labour
(32) Human trafficking
(33) Village development plans
(34) Village map
(35) Village government accountability report
(36) Level of satisfaction with village services
(37) Village data and statistics
(38) Participation of community leaders
(39) Cooperation between villages
(40) Preservation of village culture

The method of analysing the achievement of SDGs Desa is the percentage according to the existing conditions, compared to the targets of each goal in the SDGs Desa. The rate of achievement of SDGs Desa can be compiled by comparing with the target to be achieved in 2030.

The degree of the rate of achievement of SDGs Desa can be formulated as follows:

- Green: the rate will meet the 2030 SDGs Desa target
- Yellow: the rate is not expected to meet the 2030 SDGs target (+ ¼ SD)
- Red: the rate will not meet the 2030 SDGs Desa target

Chapter 25
Epilogue: After Measuring the Village

There are two moments that have strengthened the village's capacity to produce microdata. First, in 2018, for the first time, all villages collected more than 700 components of information about village conditions. This activity, summarised as the Village Development Index (IDM) data collection, has continued every year to date. Since then, a new history has been carved, in the form of village profiles and monographs, and is updated annually every July.

Second, in 2020, data collection volunteers collected a detailed list, by name and by address, of poor families and those affected by the Covid-19 pandemic. The compilation of 8.04 million microdata of these families sharpens village experience. Now the Village Data Working Group has collected *By-Name-By-Address* (BNBA) data from 106 million village residents and data from 32 million families in the villages, as well asecology data from 412 thousand Neighbourhood Units/hamlets (Rukun Tetangga/RT).

After proving capable of producing microdata, the next big step is to recognise the village as the guardian of microdata. To remain based on subsidiarity and village authority, data collection is carried out under the umbrella of a participatory paradigm for the management of collection, processing, analysis, and use of microdata.

25.1 Participatory Census

The regulation of the Minister of Villages, Development of Disadvantaged Regions, and Transmigration Number 21/2020 provides a special place for data collection as the first stage of village development and community empowerment. The next stage is planning, and implementation, as well as monitoring and evaluation of village development.

Although what is collected is numerical data, in essence, its extraction is based on the participation of village residents. The construction of data from, by, and for

A. H. Iskandar, *The Methods of Measuring the National Sustainable Development Goals in Indonesia*, Engaging Indonesia, https://doi.org/10.1007/978-981-96-0927-7_25

the village, leads to the collection of quality microdata, that is valid, accurate, and up-to-date.

The term 'microdata' here indicates direct information from the field, not analysis drawn from the district, province, or national level. All residents and all areas are recorded, asin the form of a census, not as in asurvey that takes samples.

This underlies the construction of data from the village. That is, collected data by-name-by address (BNBA) of village residents, village families, as well as details of the environmental conditions of the Neighbourhood Units/hamlets (Rukun Tetangga/RT) and village.

The construction of data by the village is shown by the work of village data volunteers from each of the Neighbourhood Units/hamlets (Rukun Tetangga/RT). They record their own neighbours. Three benefits are achieved. First, the data is accurate and valid. It is too difficult for residents to skew information, because it's hard to deceive their neighbours, who understand and are familiar with the ins and outs of their family.

Second, the data is accountable, because rechecking is carried out in Neighbourhood Units/hamlets (Rukun Tetangga/RT) meetings and in the village deliberation forum. Consensus among residents, as a characteristic of the participatory approach, is now practised to double check the truth of the information compiled by data volunteers.

Third, the data is always updated. The scope of the Neighbourhood Units/hamlets (Rukun Tetangga/RT) and village is very small. One RT is inhabited by dozens of families in remote villages, to hundreds of families in suburban villages. Therefore, the scope of the participatory census is always reachable by data volunteers. Moreover, every time there is an important event in its area or among its residents, it is always recorded to update the data. This is the basis of the activity updating village microdata throughout the year.

Participation keeps the village away from the lure of project orientation in data collection. Project orientation is characterised by discontinuous data collection and moving locations. On the contrary, participatory data collection is realised with the continuity of daily data collection, by village apparatus to the RTs, or community institutions trusted by the village government.

The processing and analysis of microdata is aimed at the interests of village residents. In addition to awakening residents who were originally incapable, living in slums, living in disaster-prone environments, data processing also supports finding potential hidden villages, the right Neighbourhood Units/hamlets (Rukun Tetangga/RT) locations for factory establishment, to supporting planning and implementation of village progress.

This internal benefit for the village encourages residents to provide honest, accurate, valid information. Because, data compilation errors are directly felt by the residents themselves, either in the form of lost opportunities, or misplacement and misuse of activities.

25.2 Smart Village

President Joko Widodo advised that the village funds and other village funding must be beneficial for all residents. This means that village development should no longer be an abstract concept, floating, not grounded. Instead, it must take the form of one-by-one village development, which is tangible on the ground, so that it is directly felt by residents and village governments. It is not an aggregate of data processing and simultaneous analysis from 75,265 villages, but an aggregate from each family, each Neighbourhood Units/hamlets (Rukun Tetangga/RT), and at the highest level becomes a village aggregate.

A micro-development strategy clearly requires microdata, which is taken directly from the field, covering all areas and all residents, and can be utilised by the village itself from planning and monitoring implementation, to evaluating the results. An example at present are the results of Village Development Index (Indeks Desa Membangun/IDM) data collection and SDGs Desa data.

After successfully becoming a data producer, the village must be recognised as a data guardian. The regulation of the Minister of Villages, Development of Disadvantaged Regions, and Transmigration Number 21/2020 has affirmed the village as the owner of microdata, including SDGs Desa data. This drives the slogan 'village data sovereignty' on the ground.

It is appropriate to position the village as a guardian of microdata. Policies are consistently based on understanding the village as a data owner. Although some of its data is stored on the ministry's server, no one is allowed to open detailed data of residents and areas other than the village administration itself. The central government, provinces, and districts, as well as the public, only have the right to open recap data.

Relations, interoperability, and consolidation of microdata with other parties should be based on the consent of the village itself. An example is the use of 'by-name-by-address' funds for social protection, government programmes, private cooperation, and empowerment by non-governmental organisations, all of which can be done only after obtaining permission from the village.

Presidential Instruction Number 4 of 2022 also applies the position of the village as a data guardian in the management of extreme poverty data. In practice, the list of extremely poor families must come from the village deliberation forum minutes, then be determined by the regent or mayor.

To broaden the benefits, microdata collection becomes a component of a smart village. There are two principles, first, village development must be based on microdata. Second, public services are supported by information technology.

The digitisation of data accompanied by the permission of the village data guardian opens up information coordination space. Village data quickly accumulates at the district, regency, province, and national levels.

Moreover, microdata collection ensures participatory census remains rooted in the locality of the area, culture, and village institutions, and the use of information technology simplifies the interoperability of village, district, province, and central data.

Appendix 1
Village Questionnaire (P.D)

P1	**Enumerator description**
P101	Name:
P102	Address:
P103	Mobile/telephone:
P2	**Location description**
P201	Province: ………………
P202	District/city: ………………
P203	Sub-district: ………………
P204	Village (and village hall photo): ………………
P205	Village email: ………………
P206	Village web address: ………………
P207	Village Facebook address: ………………
P209	Village Instagram address: ………………
P210	Village Youtube address: ………………
P211	Government status: 1. Village 2. Nagari 3. Gampong 4. Kampung 5. Sub-district
P212	Number of Sub-Village (Rukun Warga/RW) (write the number): …………
P213	Number of Neighbourhood Unit (Rukun Tetangga/RT) (write the number): …………
P214	Village establishment decree: …………
P215	Decree of the Regent/Mayor or Governor about the village map: ………………
P216	Village area size (Ha):
P217	The village is located on the island (mention the name of the island):
1.	
2.	
3.	
4.	

(continued)

A. H. Iskandar, *The Methods of Measuring the National Sustainable Development Goals in Indonesia*, Engaging Indonesia, https://doi.org/10.1007/978-981-96-0927-7

(continued)

P218	The largest topography of the village area: 1. Slope/peak 2. Valley 3. Plain
P219	Number of residents on the slope/peak (people):
P220	Village head office/village hall: 1. Available, suitable 2. Available, not suitable 3. Not available
P221	Ownership of the village head office/village hall: 1. Village asset 2. Not a village asset
P222	Location of the village head office/village hall: 1. Inside the village 2. Outside the village
P223	The main implementation of village governance is carried out at: 1. Village head office/village hall 2. Not at the village head office/village hall
P224	Working hours at the village office: 1. Uncertain/no specific schedule 2. There is a work schedule: a. starting at: …:… b. until: …:…
P225	Coordinates of the village government activities: 1. Latitude coordinates: …, … 2. Latitude: 1. North Latitude (NL) 2. South Latitude (SL) 3. East longitude line: …., …
P226	Altitude of the village government activities from sea level (DPAL): ………. metres
P227	Length of coastline (km):
P3	**Village government description**
P301	Name of the village head (and photo):
P302	National Identity Number (Nomor Induk Kependudukan/NIK) of the village head:
P303	Mobile number:
P304	Serving as village head since the year:
P305	Name of the village secretary:
P306	National Identity Number (Nomor Induk Kependudukan/NIK) of the village secretary:
P307	Mobile number:
P308	Serving as village secretary since the year:
P309	Name of the village treasurer:
P310	National Identity Number (Nomor Induk Kependudukan/NIK) of the village treasurer:
P311	Mobile number:
P312	Serving as village treasurer since the year:
P313	Name of the head of administrative affairs:
P314	National Identity Number (Nomor Induk Kependudukan/NIK) of the head of administrative affairs:
P315	Mobile number:
P316	Serving as head of administrative affairs since the year:
P317	Name of the head of financial affairs:
P318	National Identity Number (Nomor Induk Kependudukan/NIK) head of financial affairs:
P319	Mobile number:
P320	Serving as head of financial affairs since the year:
P321	Name of the head of planning:

(continued)

(continued)

P322	National Identity Number (Nomor Induk Kependudukan/NIK) head of planning:
P323	Mobile number:
P324	Serving as head of planning since the year:
P325	Name of the head of government section:
P326	National Identity Number (Nomor Induk Kependudukan/NIK) head of government section:
P327	Mobile number:
P328	Serving as head of government section since the year:
P329	Name of the head of welfare section:
P330	National Identity Number (Nomor Induk Kependudukan/NIK) head of welfare section:
P331	Mobile number:
P332	Serving as head of welfare section since the year:
P333	Name of the head of service section:
P334	National Identity Number (Nomor Induk Kependudukan/NIK) head of service section:
P335	Mobile number:
P336	Serving as head of service section since the year:
P337	Name of another village employee 1:
P338	National Identity Number (Nomor Induk Kependudukan/NIK) of another village employee 1:
P339	Mobile number:
P340	Serving as other village employee 1 since the year:
P341	Name of another village employee 2:
P342	National Identity Number (Nomor Induk Kependudukan/NIK) of another village employee 2:
P343	Mobile Number:
P344	Serving as another village official 2 since the year:
P345	Name of another village official 3:
P346	National Identity Number (Nomor Induk Kependudukan/NIK) of another village official 3:
P347	Mobile Number:
P348	Serving as another village official 3 since the year:
P349	Name of another village official 4:
P350	National Identity Number (Nomor Induk Kependudukan/NIK) of another village official 4:
P351	Mobile Number:
P352	Serving as another village official 4 since the year:
P353	Name of another village official 5:
P354	National Identity Number (Nomor Induk Kependudukan/NIK) of another village official 5:
P355	Mobile Number:

(continued)

(continued)

P356	Serving as another village official 5 since the year:
P357	Name of the Head of the Village Consultative Body (BPD):
P358	National Identity Number (Nomor Induk Kependudukan/NIK) of the Head of the Village Consultative Body (BPD):
P359	Mobile Number:
P360	Serving as the Head of the Village Consultative Body (BPD) since the year:
P361	Name of the member of the Village Consultative Body (BPD) 1:
P362	National Identity Number (Nomor Induk Kependudukan/NIK) of the member of the Village Consultative Body (BPD) 1:
P363	Mobile Number:
P364	Serving as a member of the Village Consultative Body (BPD) 1 since the year:
P365	Name of the member of the Village Consultative Body (BPD) 2:
P366	
P367	Mobile number:
P368	Serving as a member of the Village Consultative Body (BPD) 2 since the year:
P369	Name of Village Consultative Body (BPD) member 3:
P370	National Identity Number (Nomor Induk Kependudukan/NIK) of Village Consultative Body (BPD) member 3:
P371	Mobile number:
P372	Serving as a member of the Village Consultative Body (BPD) 3 since the year:
P373	Name of Village Consultative Body (BPD) member 4:
P374	National Identity Number (Nomor Induk Kependudukan/NIK) of Village Consultative Body (BPD) member 4:
P375	Mobile number:
P376	Serving as a member of the Village Consultative Body (BPD) 4 since the year:
P377	Name of Village Consultative Body (BPD) member 5:
P378	National Identity Number (Nomor Induk Kependudukan/NIK) of Village Consultative Body (BPD) member 5:
P379	Mobile number:
P380	Serving as a member of the Village Consultative Body (BPD) 5 since the year:
P381	Name of Village Consultative Body (BPD) member 6:
P382	National Identity Number (Nomor Induk Kependudukan/NIK) of Village Consultative Body (BPD) member 6:
P383	Mobile number:
P384	Serving as a member of the Village Consultative Body (BPD) 6 since the year:
P385	Name of Village Consultative Body (BPD) member 7:
P386	National Identity Number (Nomor Induk Kependudukan/NIK) of Village Consultative Body (BPD) member 7:
P387	Mobile number:
P388	Serving as a member of the Village Consultative Body (BPD) 7 since year:

(continued)

(continued)

P387	Name of the Village Consultative Body (BPD) 8 member:
P388	National Identity Number (Nomor Induk Kependudukan/NIK) of the Village Consultative Body (BPD) 8 member:
P389	Mobile number:
P390	Serving as a member of the Village Consultative Body (BPD) 8 since year:
P391	Name of the Village Consultative Body (BPD) 9 member:
P392	National Identity Number (Nomor Induk Kependudukan/NIK) of the Village Consultative Body (BPD) 9 member:
P393	Mobile number:
P394	Serving as a member of the Village Consultative Body (BPD) 9 since year:
P4	**Village consultation**
P401	Village consultation of the previous year

No	Month to (write number)	Consultation agenda	Consultation result document
1.			
2.			
3.			
4.			
5.			
6.			
7.			
8.			
9.			
10.			
11.			
12.			
13.			
14.			
15.			

P5	**Village regulations**		
P501	Previous year's Village Regulations		
No	Document number	Month (write number)	About
1.			
2.			
3.			
4.			

(continued)

(continued)

P5	Village regulations		
5.			
6.			
7.			
8.			
9.			
10.			
11.			
12.			
13.			
14.			
15.			

P502 Village Head Regulations from the Previous year

No	Document number	Month (write in numbers)	About
1.			
2.			
3.			
4.			
5.			
6.			
7.			
8.			
9.			
12.			
13.			
14.			
15.			

P503 Village Head's Decree from the Previous year

No	Document number	Month (write number)	About
1.			
2.			
3.			
4.			
5.			
6.			
7.			
8.			
9.			
10.			
11.			
12.			
13.			
14.			
15.			

P504	Village's Medium-Term Development Plan is valid until the year: ………
P505	Village's Work Plan: 1. Exists 2. Does not exist
P6	**Village budget and assets**
P601	Village's revenue budget for the previous year

Revenue Budget	IDR
1. Village Funds sourced from the National Budget	
2. Village Owned-Source Revenue (Pendapatan Asli Desa/PADes)	
3. Portion of regional tax and regional retribution results	
4. Village Fund Allocation (part of the balancing funds received by the regency/city)	
5. Financial assistance from the Provincial Budget	
6. Financial assistance from the Regency/City Budget	
7. Grants and donations from third parties	
8. Other legitimate village revenues	

P602 Village's Expenditure Budget for the Previous year

Expenditure budget	IDR
1. Field of village governance	
2. Field of village development implementation	
3. Field of community empowerment	
4. Field of community development	
5. Capital Expenditure (land, buildings, roads, bridges, computers, etc.)	
6. Capital participation in Village-Owned Enterprise (Badan Usaha Milik Desa/BUM Desa)	
7. Others (unexpected expenses, meeting consumption, etc.)	

P603 Value of Village Assets

Asset	Volume	Unit of Volume	Value in IDR
1. BUM Desa and BUMDesma		Unit	
2. Village land		Ha	
3. Village-Owned Buildings		Unit	
4. Village market		Unit	
5. Fish auction place		Unit	
6. Shops		Unit	
7. Boat mooring		Unit	
8. Public bath		Unit	

(continued)

(continued)

Asset	Volume	Unit of Volume	Value in IDR
9. Sports field		Unit	
10. Tourist location		Unit	
11. Other village assets		Unit	

P7	**Services**
P701	1. Functional computer/PC/laptop in the village head office: 1. Used 2. Rarely used 3. Not used 4. Not available 2. Internet facilities in the village head office: 1. Functional 2. Rarely functional 3. Not functional 4. Not available
P702	1. Village information system: 1. Available, updated 2. Available, not updated 3. Not available 2. Village financial system: 1. Available, updated 2. Available, not updated 3. Not available
P703	Number of poverty certificates issued in the last year (number): …….
P704	1. Number of residents who have not recorded e-KTP (people): ………… 2. Number of residents not listed in the Family Card (people): …………
P705	Village cooperation

Party involved in cooperation	Scope of cooperation (1) Between villages (2) With local government (3) With central government (4) With private sector (5) With international institutions	Year cooperation ended	Number of beneficiaries (people)	Value of cooperation (IDR)
1.				
2.				
3.				
4.				
5.				
6.				
7.				

P706	1. Village companion name: .... 2. Gender of village companion: 1. Male 2. Female 3. Village companion mobile number: ....
P8	**Village Community Institutions**
P800	Number of types of village community institutions

Name	Number of administrators (people)	Number of members (people)
1. Family Welfare Empowerment (Pemberdayaan Kesejahteraan Keluarga/PKK)		
2. Youth Organization (Karang Taruna)		
3. Customary Institution		
4. Write 1:		
5. Write 2:		
6. Write 3:		
7. Write 4:		
8. Write 5:		

P9	**Village-owned enterprises (BUM DESA)**
P901	Name of BUM Desa:
P902	Email: ....
P903	Web address: ....
P904	Facebook address: ....
P905	Twitter address: ....
P906	Village address: ....
P907	Youtube address: ....
P908	Initial capital (1) from the village government (IDR): (2) from village residents (IDR): (3) from other parties (IDR):
P909	Last year's turnover (IDR):
P910	Last year's net profit (IDR):
P911	Gross profit before tax last year (IDR):
P912	Value of BUM Desa assets (IDR):
P913	Contributions given to PADesa (IDR):
P914	Name of mentor/commissioner 1:

(continued)

(continued)

P915	National Identity Number (Nomor Induk Kependudukan/NIK):
P916	Mobile No:
P917	Name of mentor/commissioner 2:
P918	National Identity Number (Nomor Induk Kependudukan/NIK):
P919	Mobile No:
P920	Name of director/commissioner 3:
P921	National Identity Number (Nomor Induk Kependudukan/NIK):
P922	Mobile No:
P923	Name of supervisor 1:
P924	National Identity Number (Nomor Induk Kependudukan/NIK):
P925	Mobile No:
P926	Name of supervisor 2:
P927	National Identity Number (Nomor Induk Kependudukan/NIK):
P928	Mobile No:
P929	Name of supervisor 3:
P930	National Identity Number (Nomor Induk Kependudukan/NIK):
P931	Mobile No:
P932	Name of director 1:
P933	National Identity Number (Nomor Induk Kependudukan/NIK):
P934	Mobile No:
P935	Name of director 2:
P936	National Identity Number (Nomor Induk Kependudukan/NIK):
P937	Mobile No:
P938	Name of director 3:
P939	National Identity Number (Nomor Induk Kependudukan/NIK):
P940	Mobile No:
P941	BUM Desa business unit

Business unit	Number of business units	Number of Workers (people)	Net profit last year (IDR)	Turnover last year (IDR)	Business unit assets last year (IDR)
1.					
2.					
3.					
4.					
5.					
6.					
7.					

(continued)

(continued)

Business unit	Number of business units	Number of Workers (people)	Net profit last year (IDR)	Turnover last year (IDR)	Business unit assets last year (IDR)
8.					
9.					
10.					

P10	**Village infrastructure**
P1001	Transportation from the village head office to the sub-district office and the regent's office

The means of transportation commonly used by most residents from the village head office to	The means of transportation commonly used (code 1)	If there is public transportation		Travel distance (km)	Travel time (minutes)	Transportation cost (x IDR 1000)
		Type of public transport (code 2)	Main public transport (code 2)			
1. district office						
2. regent's office						
3. nearest other district office						
4. nearest other regent's office						

Code 1	Code 2
1. public transport	1. Motorcycle taxi
2. private vehicle	2. motorised vehicle with 3 or more wheels
3. bicycle, rickshaw, motorised rickshaw, horse-drawn carriage	3. boat (motorised or non-motorised)
4. walking, etc.	4. Airplane
	5. Others (rickshaw, horse-drawn carriage, ox-cart, etc.)

Appendix 2
Neighbourhood Questionnaire (P.R)

P1	**Enumerator description**
P101	Name:
P102	Address:
P103	Mobile/telephone:
P2	**Location description**
P201	Province:
P202	District/city:
P203	Subdistrict:
P204	Village:
P205	Neighbourhood Unit/Sub-village unit (RT/RW)
P206	Name of neighbourhood unit (RT) head:
P207	Address:
P208	Mobile/telephone:
P209	Neighbourhood unit (RT) location is on the island (mention the name of the island): 1. 2. 3. 4. 5.
P210	Largest topography of neighbourhood unit (RT) area: 1. slope/peak 2. Valley 3. Plain
P211	Number of residents on the slope/peak (people): ....
	Planting/maintaining trees on critical land: 1. Yes, some residents involved 2. Yes, residents not involved 3. No activity
P212	Length of coastline (km): ....
P213	Utilisation of the sea: 1. Capture fisheries: 1. Yes 2. No
P214	2. Aquaculture: 1. Yes 2. No
P215	3. Salt ponds: 1. Yes 2. No

(continued)

A. H. Iskandar, *The Methods of Measuring the National Sustainable Development Goals in Indonesia*, Engaging Indonesia, https://doi.org/10.1007/978-981-96-0927-7

(continued)

P216	4. Marine tourism: 1. Yes 2. No
P217	5. Public transportation: 1. Yes 2. No
P218	Mangrove condition: 1. Entirely good 2. Mostly good 3. Mostly bad 4. Entirely bad 5. None
P219	Mangrove planting: 1. Yes, some residents involved 2. Yes, residents not involved 3. No activity
P220	Number of residents in the coastal area (people): ....
P221	Number of residents living on water (people): ....
P222	Village area within the forest (Ha): ....
P223	Village area on the edge of the forest (Ha): ....
P224	Forest functions: 1. conservation (Ha): .... 2. protection (Ha): .... 3. production (Ha): .... 4. village forest (Ha): ....
P225	Number of residents living within the forest (people): ....
P226	Number of residents living around the forest (people): ....
P227	Residents' dependence on the forest: 1. High 2. Medium 3. Low 4. Not dependent
P228	Forest reforestation: 1. Yes, some residents involved 2. Yes, residents not involved 3. No activity
P3	**Description Of Sub-village (RW) Management**
P301	Name of Sub-village (RW) Chairman (and photo):
P302	National Identity Number (Nomor Induk Kependudukan/NIK) of Sub-village (RW) Chairman:
P303	Mobile/telephone:
P304	Serving as Sub-village (RW) Chairman since the year:
P305	Name of Sub-village (RW) Secretary:
P306	National Identity Number (Nomor Induk Kependudukan/NIK) of Sub-village (RW) Secretary:
P307	Mobile/telephone:
P308	Serving as Sub-village (RW) Secretary since the year:
P309	Name of Sub-village (RW) Treasurer:
P310	National Identity Number (Nomor Induk Kependudukan/NIK) of Sub-village (RW) Treasurer:
P311	Mobile/telephone:
P312	Serving as Sub-village (RW) Treasurer since the year:
P4	**Neighbourhood Unit (RT) management description**
P401	Name of Neighbourhood Unit (RT) Chairman (and photo):
P402	National Identity Number (Nomor Induk Kependudukan/NIK) of Neighbourhood Unit (RT) Chairman:
P403	Mobile/telephone:

(continued)

(continued)

P404	Serving as Neighbourhood Unit (RT) Chairman since the year:
P405	Name of Neighbourhood Unit (RT) Secretary:
P406	National Identity Number (Nomor Induk Kependudukan/NIK) of Neighbourhood Unit (RT) Secretary:
P407	Mobile/telephone:
P408	Serving as Neighbourhood Unit (RT) Secretary since the year:
P409	Name of Neighbourhood Unit (RT) Treasurer:
P410	National Identity Number (Nomor Induk Kependudukan/NIK) of Neighbourhood Unit (RT) Treasurer:
P411	Mobile/telephone:
P412	Serving as Neighbourhood Unit (RT) Treasurer since the year:
P5	**Economic institutions**
P501	Agency for deploying TKI abroad: (1) Number (companies): (2) Number (people):
P502	Industry by main raw material

Type of industry	Number of small and household industries (workers under 20 people)	Number of medium and large industries	Number of management (people)	Number of workers (people)
1. Leather goods industry (bags, shoes, sandals, etc.)				
2. Wood goods industry (tables, chairs, cabinets, etc.)				
3. Precious metal or metal goods industry (furniture, jewellery, etc.)				
4. Heavy metal industry				
5. Cloth goods industry (weaving, convection, etc.)				

(continued)

(continued)

Type of industry	Number of small and household industries (workers under 20 people)	Number of medium and large industries	Number of management (people)	Number of workers (people)
6. Pottery/ceramic/ stone industry (porcelain, ceramics, tiles, etc.)				
7. Tile and brick industry				
8. Wicker industry from rattan/ bamboo/grass/ pandan, etc.				
9. Food and beverage industry (processing and preservation of meat, fish, fruits, oils and fats, milk and dairy products, etc.)				
10. Tobacco industry				
11. Other industry 1, write: …………….				
12. Other industry 2, write: …………….				
13. other industries 3, write: …………….				

P503	Industrial layout; 1. Number of industrial centres: …………… 2. Number of Small Industrial Areas (Lingkungan Industri Kecil/LIK): …………… 3. Number of Small Industrial Villages ((Perkampungan Industri Kecil/PIK): ……………
P504	Presence of pub/disco/karaoke place: 1. Yes 2. No, nearest ……. km

(continued)

(continued)

P505	Kerosene and Liquid Petroleum Gas (LPG) base: 1. Presence of kerosene base/agent/seller (including roaming kerosene sellers): 1. Yes 2. No 2. Presence of Liquid Petroleum Gas (LPG) base/agent/seller (stalls, shops, supermarkets, roaming gas sellers): 1. Yes 2. No
P506	Presence of cooperatives: 1. Number of Village Unit Cooperative (Koperasi Unit Desa/KUD): .... 2. Number of Village Unit Cooperative (Koperasi Unit Desa/KUD) buying/selling agricultural products/production: .... 3. Number of Village Unit Cooperative (Koperasi Unit Desa/KUD) providing Business Credit: .... 4. Number of Village Unit Cooperative (Koperasi Unit Desa/KUD) conducting other activities: .... 5. Number of Small Industry and People's Craft Cooperatives (Kopinkra)/Micro businesses: .... 6. Number of Savings and Loan Cooperatives (Kospin): .... 7. Number of multi-purpose cooperatives: .... 8. Number of other cooperatives: ....
P507	Farmer/fisherman production facilities kiosks: 1. Owned by Village Unit Cooperative (Koperasi Unit Desa/KUD) (unit): .... 2. Owned by Village-owned Enterprise (Badan Usaha Milik Desa/BUM Desa) (unit): .... 3. Owned by others than Village Unit Cooperative (Koperasi Unit Desa/KUD) and Village-owned Enterprise (BUM Desa): ....
P508	Available economic facilities

Type	Number	Condition: 1. Good 2. Bad 3. Not functioning 4. Not available	If Not Available, distance to nearest facility (km)	Ease of reaching: 1. Very easy 2. Easy 3. Difficult 4. Very difficult
1. Shopping group (minimum 10 shops and grouped in one location)				
2. Market with permanent building (has a roof, floor, and walls)				
3. Market with semi-permanent building (has a roof and floor, without walls)				

(continued)

(continued)

Type	Number	Condition: 1. Good 2. Bad 3. Not functioning 4. Not available	If Not Available, distance to nearest facility (km)	Ease of reaching: 1. Very easy 2. Easy 3. Difficult 4. Very difficult
4. Market without a building (for example: dawn market, floating market, etc.)				
5. Number of minimarkets/ supermarkets (business places in permanent buildings for selling various types of goods retail with price tags, self-service system, floor area < 400 m^2)				
6. Shop/grocery store (business places in permanent buildings for selling various types of daily necessities retail, without a self-service system)				
7. Shop/grocery store selling food ingredients (staples)				
8. Restaurant (ready-to-eat food business in a permanent building, customers are usually taxed)				

(continued)

(continued)

Type	Number	Condition: 1. Good 2. Bad 3. Not functioning 4. Not available	If Not Available, distance to nearest facility (km)	Ease of reaching: 1. Very easy 2. Easy 3. Difficult 4. Very difficult
9. Food and drink stall (ready-to-eat food business in a permanent building, customers are usually not taxed)				
10. Hotel (provides accommodation services and has a restaurant, lodging with a business license as a hotel)				
11. Lodging: hostel/ motel/ guesthouse/ boarding house (provides accommodation, lodging with a business license not as a hotel)				
12. Car/motorcycle workshop				
13. Beauty Salon				
14. Ticket/Travel/ Travel Agency				
15. BRI Bank				
16. BRI Bank Agent				
17. BNI Bank				
18. BNI Bank Agent				
19. Mandiri Bank				
20. Mandiri Bank Agent				
21. BPD				
22. BPD Bank Agent				
23. Other public general banks				
24. BNI Bank				
25. BNI Bank Agent				

(continued)

(continued)

Type	Number	Condition: 1. Good 2. Bad 3. Not functioning 4. Not available	If Not Available, distance to nearest facility (km)	Ease of reaching: 1. Very easy 2. Easy 3. Difficult 4. Very difficult
26. Mandiri Bank				
27. Mandiri Bank Agent				
28. BCA Bank				
29. CIMB-Niaga/ Maybank Bank				
30. Sinarmas Bank				
31. Permata Bank				
32. Other private banks				
33. BPR (People's Credit Bank)				
34. Baitul Maal Wa Tamwil (BMT)				
35. Pawnshop				
36. Automatic Teller Machine (ATM)				
37. Other economic facilities				

P509	Credit facilities 1. People's Business Credit (Kredit Usaha Rakyat/KUR): 1. Yes 2. No 2. Food and Energy Resilience Credit (Kredit Ketahanan Pangan dan Energi/KKP-E): 1. Yes 2. No 3. Small Business Credit (Kredit Usaha Kecil/KUK): 1. Yes 2. No 4. Joint Business Group (Kelompok Usaha Bersama/KUBE): 1. Yes 2. No
P6	**Infrastructure**
P601	Main road lighting 1. Electricity provided by the government 2. Electricity provided by non-government 3. Non-electricity 4. None
P602	Inter-Neighbourhood Unit (RT) transportation infrastructure: 1. Land 2. Water 3. Land and water 4. Air
P603	Road length (km) 1. Asphalt road: ….. 4. Boardwalk over water: …. 2. Hardened road (gravel, stone, etc.): ….. 5. Footpath: …. 3. Dirt road: ….. 6. Other roads: ….
P604	Land roads can be passed by 4-wheeled motor vehicles or more: 1. all year round 3. during the dry season 2. all year round except at certain times (when it rains, high tide, etc.) 4. cannot be passed all year round

(continued)

(continued)

P605	Presence of public transportation 1. Public transportation route: 1. Fixed route 2. Without a fixed route 3. No public transportation 2. Public transportation operation: 1. Every day 2. Not every day 3. Public transportation operating hours: 1. Day and night 4. Daytime only
P606	Sea/river dock 1. Exists, good condition 2. Exists, poor condition 3. Exists, non-functional 4. Does not exist
P607	Mobile signal 1. Number of Base Transceiver Station (BTS) towers: .... 2. Mobile phone service operators covering the area:

Operator	Signal in most of the area: 1. Very strong signal 2. Strong signal 3. Weak signal 4. No signal	Internet signal: 1. 4G/LTE 2. 3G/H/H + /EVDO 3. 2.5G/E/GPRS 4. No internet signal
1. Telkomsel/Halo/Loop/As		
2. Indosat		
3. XL/Axis		
4. Hutchison 3		
5. PSN ByRU		
6. Smartfren		
7. Bakrie Telecom		

P608	1. Post office/assistant post office/post house: 1. Operating 2. Rarely operating 3. Not operating 4. Not available 2. Mobile post service: 1. Available 2. Not available 3. Private courier service company/agent (for sending goods/documents): 1. Operating 2. Rarely operating 3. Not operating 4. Not available
P609	TV/radio programmes/broadcasts received

TV and radio programmes/ broadcasts	Can it be received? 1. Yes 2. No	If the programme/broadcast can be received, is it necessary to use a satellite dish/cable TV? 1. Yes 2. No
1. TVRI		
2. Regional TVRI		
3. Private TV		
4. Foreign TV		

(continued)

(continued)

TV and radio programmes/ broadcasts	Can it be received? 1. Yes 2. No	If the programme/broadcast can be received, is it necessary to use a satellite dish/cable TV? 1. Yes 2. No
5. RRI		
6. Regional RRI		
7. Private radio		
8. Community radio		

P610	1. Number of illegal settlement locations: ………… 2. Number of public facilities/social facilities occupied by residents: 1. Market: ……… 4. Under the bridge: ……… 2. Station: ……… 5. Port: ……… 3. Terminal: ………
P611	Number of special settlement locations (1) luxury settlements/housing: ……… (2) apartments: ……… (3) flats: ……… (4) boarding schools: ……… (5) boarding houses: ……… (6) barracks/military dormitories (7) prisons/detention centres: ………
P7	**Environment and natural disasters**
P701	Land area by type of land use (Ha) (1) Irrigated rice fields: ………………… (2) Non-irrigated rice fields (rainfed, tidal, swamp, etc.): ………………… (3) Gardens: ………………… (4) Shifting cultivation: ………………… (5) Fish ponds: ………………… (6) Ponds/ditches/ponds: ………………… (7) Pasture land: ………………… (8) Plantation company land: ………………… (9) Forest area: ………………… (10) Other non-rice agricultural land: ………………… (11) Mining land: ………………… (12) Residential land: ………………… (13) Office land: ………………… (14) Commercial land: ………………… (15) Industrial land: ………………… (16) Public facilities (fields, public infrastructure, roads, docks, etc.): ………………… (17) Other land: …………………
P702	Name of the river crossing: 1. ……………………………….. 2. ……………………………….. 3. ……………………………….
P703	Name of lake/reservoir/pond: 1. ……………………………….. 2. ……………………………….. 3. ……………………………….

(continued)

(continued)

P704	Number of springs: ....
P705	Number of small reservoirs: ....
P706	Use of rivers, irrigation channels, lakes/reservoirs/dams, and small reservoirs

No	Type of Use	River		Irrigation Channel		Lake/reservoir/pond/dam		Small Reservoir	
		1. Used 2. Not used 3. Not available	1. Good Condition 2. Polluted 3. Not available	1. Used 2. Not used 3. Not available	1. Good Condition 2. Damaged 3. Not available	1. Used 2. Not used 3. Not available	1. Good Condition 2. Polluted 3. Not available	1. Used 2. Not used 3. Not available	1. Good Condition 2. Damaged 3. Not available
1.	Bathing/ washing								
2.	Source of drinking/ cooking water								
3.	Raw material for drinking water								
4.	Irrigation of agricultural land								
5.	Tourism (commercial)								
6.	Fisheries								
7.	Transportation								
8.	Power generation								
9.	Industry/ factory								
10.	Others								

P707	If there is pollution in the river/reservoir/lake/dam: 1. Waste source from: a. Factory/industry/business: 1. Yes 2. No b. Household: 1. Yes 2. No c. Others: 1. Yes 2. No 2. The source of waste is located at: 1. Within Neighbourhood Unit (RT) 2. Within Sub-village unit (RW) 3. Within the village 4. Outside the village
P709	Pollution over the past year

Environment	Pollution: 1. Present 2. Not present	Source of pollution: 1. Factory/industry/business 2. Household 3. Others	Location of waste: 1. Within Neighbourhood Unit (RT) 2. Within Within Sub-village unit (RW) 3. Within the village 4. Outside the village	Residents' complaints 1. Present 2. Not present
1. Water				
2. Soil				
3. Air				

P710	Waste recycling: 1. Present, some residents involved 2. Present, residents not involved 3. No activity
P711	The community's habit of burning fields/gardens in the village for agricultural processes: 1. Present 2. Not present
P712	Presence of Group C excavation sites (for example: river stones, sand, limestone, kaolin, quartz sand, clay, etc.): 1. Present 2. Not present
P713	Natural disasters in the past year

Natural disasters	Occurrence: 1. Yes 2. No	Number of occurrences (quantity)	Casualties (people)	Refugees (people)	Affected residents (people)
1. Landslide					
2. Flood					
3. Flash flood					
4. Earthquake					
5. Tsunami					
6. Sea market wave					
7. Whirlwind/tornado/typhoon					

(continued)

(continued)

Natural disasters	Occurrence: 1. Yes 2. No	Number of occurrences (quantity)	Casualties (people)	Refugees (people)	Affected residents (people)
8. Volcanic eruption					
9. Forest/land fire					
10. Drought (land)					

P714	Facilities/efforts for anticipation/mitigation of natural disasters: 1. Early warning system for natural disasters: 1. Yes 2. No 2. Specific early warning system for tsunamis: 1. Not a potential area tsunami 2. Yes 3. No 3. Safety equipment (rubber boat, tent, mask, etc.): 1. Yes 2. No 4. Disaster evacuation signs and routes: 1. Yes 2. No 5. Creation/maintenance/normalisation: river, canal, embankment, ditch, drainage, reservoir, beach, etc.: 1. Yes 2. No
P8	**Education**
P801	Presence of educational facilities

Level	Name	Owner: 1. Public 2. Private	Building condition 1. Suitable 2. Damaged	Teachers (persons)	Students (persons)	Other staff (persons)
1. Early Childhood Education Post (Pos PAUD)	1.					
	2.					
	3.					
	4.					
	5.					
2. Kindergarten (Taman Kanak-Kanak)	1.					
	2.					
	3.					
	4.					
	5.					
3. Primary School/ Elementary School	1.					
	2.					
	3.					
	4.					
	5.					

(continued)

(continued)

Level	Name	Owner: 1. Public 2. Private	Building condition 1. Suitable 2. Damaged	Teachers (persons)	Students (persons)	Other staff (persons)
4. Special Needs Primary School	1.					
	2.					
	3.					
5. Junior High School	1.					
	2.					
	3.					
6. Special Needs Junior High School	1.					
	2.					
	3.					
7. Senior High School	1.					
	2.					
	3.					
8. Senior Vocational High School	1.					
	2.					
	3.					
9. Special Needs Senior High School	1.					
	2.					
	3.					
10. Academy/ Higher Education Institution	1.					
	2.					
	3.					
11. Islamic Boarding School	1.					
	2.					
	3.					
12. Religious Elementary School	1.					
	2.					
	3.					
13. Seminary/ similar	1.					
	2.					
	3.					
14. Other religious schools	1.					
	2.					
	3.					

(continued)

(continued)

Level	Name	Owner: 1. Public 2. Private	Building condition 1. Suitable 2. Damaged	Teachers (persons)	Students (persons)	Other staff (persons)
15. Activities for the eradication of illiteracy/functional literacy	1.					
	2.					
	3.					
16. Catch-up Package A Activities	1.					
	2.					
	3.					
17. Catch-up Package B Activities	1.					
	2.					
	3.					
18. Catch-up Package C Activities	1.					
	2.					
	3.					
19. Playgroup/play group	1.					
	2.					
	3.					
20. Childcare centre	1.					
	2.					
	3.					
21. Quran Education Park	1.					
	2.					
	3.					
22. Foreign language course	1.					
	2.					
	3.					
23. Computer course	1.					
	2.					
	3.					
24. Sewing/fashion design course	1.					
	2.					
	3.					

(continued)

(continued)

Level	Name	Owner: 1. Public 2. Private	Building condition 1. Suitable 2. Damaged	Teachers (persons)	Students (persons)	Other staff (persons)
25. Beauty course	1.					
	2.					
	3.					
26. Car/ motorcycle mechanic course	1.					
	2.					
	3.					
27. Driving course	1.					
	2.					
	3.					
28. Electronics course	1.					
	2.					
	3.					
29. Cooking/ culinary arts course	1.					
	2.					
	3.					
30. Typing course	1.					
	2.					
	3.					
31. Accounting course	1.					
	2.					
	3.					
32. Other courses	1.					
	2.					
	3.					

P802	Library or reading park: 1. Available 2. Not available
P9	**Health**
P901	Presence of health facilities

Level	Name	Owner: 1. State 2. Private	Doctor (person)	Midwife (person)	Health Workers (person)	Other Staff (person)
1. Hospital	1.					
	2.					

(continued)

(continued)

Level	Name	Owner: 1. State 2. Private	Doctor (person)	Midwife (person)	Health Workers (person)	Other Staff (person)
	3.					
2. Maternity hospital	1.					
	2.					
	3.					
3. Health centre with inpatient care	1.					
4. Health centre without inpatient care	1.					
5. Auxiliary health centre	1.					
	2.					
3						
6. Polyclinic/ treatment centre	1.					
	2.					
	3.					
7. Doctor's practice	1.					
	2.					
	3.					
8. Maternity home	1.					
	2.					
	3.					
9. Midwife's practice	1.					
	2.					
	3.					
10. Village health post	1.					
	2.					
	3.					
11. Village Maternity Clinic	1.					
	2.					
	3.					
12. Pharmacy	1.					
	2.					
	3.					

(continued)

(continued)

Level	Name	Owner: 1. State 2. Private	Doctor (person)	Midwife (person)	Health Workers (person)	Other Staff (person)
13. Specialised Medicine/Herbal Medicine Shop	1.					
	2.					
	3.					
14. Integrated Service Post	1.					
	2.					
	3.					
15. Integrated Guidance Post (Pos Binaan Terpadu/ Posbindu)	1.					
	2.					
	3.					
16. Baby shaman/ midwife/paraji practice place	1.					
	2.					
	3.					

P902 Extraordinary Events and Diseases in the Last year

Type	Event: 1. Yes 2. No	Number of sufferers	Number of deaths
1. Gastroenteritis/diarrhoea			
2. Dengue fever			
3. Measles			
4. Malaria			
5. Bird flu/SARS			
6. Covid-19			
7. Hepatitis B			
8. Hepatitis E			
9. Diphtheria			
10. Chikungunya			
11. Leptospirosis			
12. Cholera			
13. Malnutrition (marasmus and kwashiorkor)			
14. Others			

P10	**Social Cultural Religion**
P1001	Number of residents participating in health social security (people): …………
P1002	Number of residents participating in employment social security (people): …………
P1003	Places of worship (number) 1. Mosque: ……… 6. Pura: ……… 2. Musala/surau/langgar: ……… 7. Vihara: ……… 3. Christian Church: ……… 8. Temple: ……… 4. Catholic Church: ……… 9. Temple: ……… 5. Chapel: ……… 10. Others: ………
P1004	Presence of religious institutions

Name	Number of administrators (people)	Number of members (people)	Facilities 1. Available, good 2. Available, moderately damaged 3. Yes, severely damaged 4. None
1.			
2.			
3.			
4.			
5.			
6.			
7.			
8.			
9.			
10.			

P1005	Cultural heritage site (specify) 1. ………………………… 2. ………………………… 3. …………………………
P1006	Presence of isolated tribes (1) estimated number of families: ……… (2) estimated population: ………
P1007	Open public spaces primarily intended as a place for villagers/town residents to relax/play without having to pay (for example: open fields/squares, parks, etc.): 1. Yes, managed 2. Yes, not managed 3. None

(continued)

(continued)

P1008	Name of local wisdom/customs (write down): 1. pregnancy:…………………………………………… 5. Marriage:…………………………………………… 2. birth:…………………………………………… 6. Community life:…………………………………………… 3. Occupation/livelihood:……………………………… 7. Death:…………………………………………… 4. Environment:………………………………
P1009	Number of types of village community institutions

Name	Number of groups/ institutions	Number of administrators (people)	Number of members (people)	Facilities 1. Available, good 2. Available, moderately damaged 3. Available, severely damaged 4. Not available
1. Farmer group				
2. Community				
3. Water management				
4. Music				
5. Traditional/ethnic music				
6. Dance				
7. Traditional/ethnic dance				
8. Painting				
9. Drama				
10. Photography				
11. Other arts				
12. Football				
13. Volleyball				
14. Badminton				
15. Basketball				
16. Lawn tennis				
17. Table tennis				
18. Futsal				
19. Swimming				
20. Martial arts				
21. Billiards				

(continued)

(continued)

Name	Number of groups/ institutions	Number of administrators (people)	Number of members (people)	Facilities 1. Available, good 2. Available, moderately damaged 3. Available, severely damaged 4. Not available
22. Fitness/aerobics				
23. Other sports				

P11	Security
P1101	Incidents of mass fighting in the last year

Type	Main cause (CODE)	Number of incidents	Injured victims (lives)	Fatal victims (lives)	Resolution 1. Yes, all 2. Yes, partially 3. No	Main mediator party (CODE)
1. between community groups						
2. community groups between villages						
3. community groups with security forces						
4. community groups with government officials						
5. between security forces						
6. between government officials						
5. students						
6. between tribes						
7. others						

Main Cause Code	Main Mediator Code
(1) Wealth (2) Power (3) Romance (4) Ideology (5) Religion/belief (6) Crowds (sports, entertainment, etc.) (7) Dissatisfaction with policy/service (8) Others	(1) Security forces (2) Village government officials (3) Regional government officials (4) Community leaders (5) Religious leaders (6) Others (7) None

P1102 Crimes that Occurred in the Village Over the Past year

Type of crime	Number of cases	Number of cases resolved	Number of cases that could not be handled	Victims injured (people)	Victims killed (people)
1. Theft					
2. Robbery					
3. Fraud/embezzlement					
4. Assault					
5. Arson					
6. Rape/crimes against decency					
7. Drug abuse/trafficking					
Gambling					
Murder					
Human trafficking					
Corruption					
Others					

P1103	Community activities to maintain neighbourhood security over the past year: 1. Number of security post construction/maintenance activities: .... 2. Number of security team formation/arrangement activities: .... 3. Number of added civil defence/village security members (people): .... d. Reporting guests who stay for more than 24 h to local authorities: 1. Yes 2. No e. Activation of neighbourhood security system from citizen initiative: 1. Yes 2. No
P1104	Number of civil defence/village security members (people): ....

(continued)

(continued)

P1105	1. Number of police posts (including police stations): 1. in use (units): …… 2. not in use (units): …… 2. If there is no police post, estimated distance to the nearest police post (including police station) (km): ……. 3. Ease of reaching the nearest police post (including police station): 1. Very easy 2. Easy 3. Difficult
P1106	Number of suicide victims (including attempted suicides) that occurred over the past year (people): ……….
P1107	1. Number of gathering locations for street children (other than shelter homes): ……. 2. Number of gathering places for homeless/beggars: ………
P1108	Number of red-light districts/gathering places for Commercial Sex Workers (CSW): …………

Appendix 3
Family Questionnaire (P.K)

P1	**Enumerator description**
P101	Name:
P102	Address:
P103	Mobile/telephone:
P2	**Location description**
P201	Province:
P202	District/city:
P203	Sub-district:
P204	Village:
P205	Neighbourhood Unit/Sub-village unit (RT/RW)
P206	Name:
P207	Address:
P208	Mobile number:
P209	Landline/home phone number:
P3	**Family description**
P301	Family card number:
P302	Head of family's National Identification Number (Nomor Induk Kependudukan/NIK):
P4	**Settlement**
P401	Occupied dwelling place: 1. Owned 2. Contract/Rent 3. Free Rent 4. Borrowed 5. Service 6. Other
P402	Status of the land of the occupied dwelling place: 1. Owned 2. Owned by others 3. State land 4. Other
P403	1. Floor area of dwelling place: ….. (m^2) 2. Land area of dwelling place: ….. (m^2)

(continued)

A. H. Iskandar, *The Methods of Measuring the National Sustainable Development Goals in Indonesia*, Engaging Indonesia, https://doi.org/10.1007/978-981-96-0927-7

(continued)

P404	Type of the most extensive floor in the dwelling place: 1. Marble/granite 5. High-quality wood/board 9. bamboo 2. Ceramic 6. Cement/red brick 10. Other 3. Parquet/vinyl/carpet 7. Bamboo 4. Tile/terrazzo 8. Low-quality wood/board
P405	The majority of the house walls: 1. Cement/concrete/high-quality wood 2. Low-quality wood/bamboo 3. Other
P406	Windows: 1. Present, functional 2. Present, non-functional 3. Absent
P407	Roof: 1. Tiles 2. Wood/Straw 3. Other
P408	House lighting: 1. PLN electricity 4. Other light sources 2. Non-PLN electricity 5. None 3. Oil lamp/candle
P409	Energy for cooking: 1. City gas/LPG/biogas (to P407) 3. Firewood 2. Kerosene/coal (to P407) 4. Other (to P407)
P410	If using firewood for cooking, source of firewood: 1. Purchase 2. Taken from the forest 3. Taken outside/non-forest 4. other
P411	Waste disposal place: 1. None 2. In garden/river/drainage 3. Burned 4. Rubbish bin 5. Regularly collected rubbish bin
P412	Toilet facilities: 1. Private 2. Shared/neighbour's 3. Public toilet 4. None
P413	Main source of bathing water: 1. Paid piped water/refillable water/bottled water 4. River, lake, reservoir 2. Piped water 5. Rainwater harvesting 3. Spring/well 6. Others
P414	Toilet facilities: 1. Private toilet 3. Public toilet 2. Shared/neighbour's toilet 4. Others
P415	Main source of drinking water: 1. Paid piped water/refillable water/bottled water 4. Rainwater harvesting 2. Spring/piped water/well 5. Others 3. River, lake, reservoir
P416	Liquid waste disposal site: 1. Tank/waste management installation 3. Hole in the ground 2. Rice field/pond/river/drainage/sea 4. Others
P417	House located under high voltage power lines: 1. Yes 2. No
P418	House on the riverbank: 1. Yes 2. No
P419	House on a hill/mountain slope: 1. Yes 2. No
P420	Overall condition of the house: 1. Slum 2. Not slum
P421	Nearest Education Access

No	Facility	Distance (km)	Travel time (hours)	Accessibility 1. Easy 2. Difficult
1.	Early Childhood Education (PAUD)			
2.	Kindergarten/Islamic Kindergarten (TK/RA)			
3.	Primary School or equivalent			
4.	Junior High School or equivalent			
5.	High School or equivalent			
6.	Higher Education			
7.	Islamic Boarding School			
8.	Seminary			
9.	Other religious education			

P422 Access to Nearest Health Facility

No	Facility	Distance (km)	Travel tme (hours)	Ease 1. Easy 2. Difficult
1.	Hospital			
2.	Maternity Hospital			
3.	Polyclinic			
4.	Community Health Centre			
	Auxiliary Community Health Centre			
	Village Maternity Post			
	Village Health Post			
	Integrated Service Post			
	Pharmacy			
	Drugstore			

P423 Access to Nearest Health Workers

No	Facility	Distance (km)	Travel time (hours)	Ease 1. Easy 2. Difficult
	Specialist doctor			
	General practitioner			

(continued)

(continued)

No	Facility	Distance (km)	Travel time (hours)	Ease 1. Easy 2. Difficult
	Midwife			
	Health worker			
	Shaman			

P424 Access to Transportation Infrastructure and Facilities

No	Destination	Oldest type of transportation: 1. Land 2. Water 3. Air	Use of public transportation 1. Yes 2. No	One-way travel time (hours)	One-way cost (IDR)	Ease: 1. easy 2. difficult
1	Main job location					
2	Currently cultivated farmland					
3	School					
4	Medical treatment					
5	Weekly/monthly/yearly worship					
6	Nearest recreation					

P425	Beneficiary/recipient of government programs 1. BLT Village Funds: 1. Yes 2. No 6. Assistance for workers 1. Yes 2. No 2. Family Hope Program/PKH: 1. Yes 2. No 7. Child education assistance 1. Yes 2. No 3. Cash Social Assistance/BST: 1. Yes 2. No 8. Others 1. Yes 2. No 4. Presidential Assistance/Banpres: 1. Yes 2. No 5. UMKM Assistance: 1. Yes 2. No

Appendix 4
Individual Questionnaire (P.I)

P1	**Individual description**
P101	Family Card Number:
P102	National Identity Number (Nomor Induk Kependudukan/NIK):
P103	Name:
P104	Gender: 1. Male 2. Female
P105	Place of birth:
P106	Date of birth:
P107	Marital status: 1. Married 2. Not married 3. Widowed/divorced
P108	Religion: 1. Islam 2. Christian 3. Catholic 4
P109	Ethnic group (write down):
P110	Citizenship: 1. Indonesian citizen 2. Foreign citizen
P111	Mobile number:
P112	WhatsApp number:
P113	Personal email address:
P114	Personal Facebook address:
P115	Personal Twitter address:
P116	Personal Instagram address:
P2	**Job description**

(continued)

A. H. Iskandar, *The Methods of Measuring the National Sustainable Development Goals in Indonesia*, Engaging Indonesia, https://doi.org/10.1007/978-981-96-0927-7

(continued)

P201	Employment status: 1. in school 2. Housewife 3. Unemployed 4. Job seeking 5. Employed
P202	Main job: 1. Landowner farmer 2. Tenant farmer 3. Farm labourer 4. Boat/ship owner fishermen 5. Boat/ship renter fishermen 6. Fishermen labourers 7. Teacher 8. Religious teacher 9. Trader 10. Processing/industry 11. Civil servant 12. Military personnel 13. Village apparatus 14. Village office employee 15. Indonesian migrant worker 16. Others
P203	1. Social security for workers from the government: 1. Participant 2. Non-participant 1. Social security for workers from private sector: 1. Participant 2. Non- participant
P204	Last year's income from (IDR):

No	Source of Income	Volume		Annual Income (IDR)	Exported (1) All (2) Most of it (3) No
		Quantity	Unit		
1	Rice		ton		
2	Secondary crops (corn, legumes, tubers, etc.)		ton		
3	Horticulture (fruits, vegetables, ornamental plants, medicinal plants, etc.)		kg		
4	Rubber		ton		
5	Palm oil		ton		
6	Coffee		kg		
7	Cocoa		ton		
8	Coconut		ton		
9	Pepper		kg		
10	Cloves		kg		
11	Tobacco		kg		
12	Sugarcane		Ton		
13	Beef cattle		head		
14	Cow's milk		Litre		
15	Sheep		Head		
16	Other large livestock (horses, buffalo, etc.)		Head		
17	Broiler chicken		Head		
18	Chicken eggs		Kg		

(continued)

(continued)

No	Source of Income	Volume		Annual Income (IDR)	Exported (1) All (2) Most of it (3) No
		Quantity	Unit		
19	Other small livestock (ducks, birds, etc.)		Head		
20	Capture fisheries (including other biota)		Kg		
21	Aquaculture (including other biota)		Kg		
22	Bamboo		Stem		
23	Forestry plant cultivation (teak, mahogany, sengon, etc.)		Stem		
24	Forest product collection (honey, agarwood, fruits, firewood, rattan, etc.)		kg		
25	Wild animal capture (pigs, jungle fowl, deer, etc.)		Head		
26	Wild animal breeding (arowana, crocodile, etc.)		Head		
27	Agricultural services (tractor rental, milling, etc.)		Day		
28	Mining and quarrying		kg		
29	Craft industry		Month		
30	Processing industry		Month		
31	Trade		Month		
32	Stalls and restaurants		Month		
33	Transport		Month		
34	Warehousing		Month		
35	Communication		Month		
36	Services outside agriculture		Month		
37	Others		Month		
38	Permanent employees		Month		
39	Temporary employees		Month		
40	TNI (Indonesian National Armed Forces)		Month		
41	Civil Servants		Month		
42	Indonesian Migrant Workers abroad		Month		

(continued)

(continued)

No	Source of Income	Volume		Annual Income (IDR)	Exported (1) All (2) Most of it (3) No
		Quantity	Unit		
43	Contributions (from family, from government)		Month		
44	Others		Month		

P4	**Health description**
P401	Diseases suffered in the last year 1. Gastroenteritis/diarrhoea: 1. Yes 2. No 11. Leptospirosis: 1. Yes 2. No 2. Dengue fever: 1. Yes 2. No 12. Cholera: 1. Yes 2. No 3. Measles: 1. Yes 2. No 13. Malnutrition (marasmus and kwashiorkor): 1. Yes 2. No 4. Malaria: 1. Yes 2. No 14. Heart disease: 1. Yes 2. No 5. Bird flu/SARS: 1. Yes 2. No 15. Pulmonary TB: 1. Yes 2. No 6. Covid-19: 1. Yes 2. No 16. Cancer: 1. Yes 2. No 7. Hepatitis B: 1. Yes 2. No 17. Diabetes: 1. Yes 2. No 8. Hepatitis E: 1. Yes 2. No 18. Paralysis: 1. Yes 2. No 9. Diphtheria: 1. Yes 2. No 19. Other: 1. Yes 2. No 10. Chikungunya: 1. Yes 2. No
P402	How many times the following health facilities were visited in the last year for treatment/care (number) 1. Hospital: ...... 9. Midwife practice place: ...... 2. Maternity hospital: ...... 10. Poskesdes: ...... 3. Health centre with inpatient care: ...... 11. Polindes: ...... 4. Health centre without inpatient care: ...... 12. Pharmacy: ...... 5. Auxiliary health centre: ...... 13. Special medicine/herbal store: ...... 6. Polyclinic/treatment centre: ...... 14. Posyandu: ...... 7. Doctor's practice place: ...... 15. Posbindu: ...... 8. Maternity home: ...... 16. Traditional birth attendant practice place/paraji: ......
P403	1. Government health insurance: 1. Participant 2. Non-participant 2. Private health insurance: 1. Participant 2. Non-participant
P404	Disabilities 1. Blindness: 1. Yes 2. No 2. Deafness: 1. Yes 2. No 3. Muteness: 1. Yes 2. No 4. Deaf-mute: 1. Yes 2. No 5. Physical disability: paralysis/abnormality/incomplete limbs: 1. Yes 2. No 6. Intellectual Disability (mental disability, mental retardation): 1. Yes 2. No 7. Emotional Disability (ex-mental illness, difficulty controlling emotions and social control): 1. Yes 2. No 8. Disability ex-leprosy: had leprosy and declared cured by a doctor: 1. Yes 2. No 9. Multiple Disabilities (physical-mental): physical disability and mental disability: 1. Yes 2. No 10. Shackled: 1. Yes 2. No
P5	**Education description**

(continued)

(continued)

P501	Highest education completed: 1. No schooling 5. Diploma 1–3 9. Religious school, seminary, 2. Primary school and equivalent 6. Bachelor's degree and equivalent monastery and the like 3. Junior high school and equivalent 7. Master's degree and equivalent 10. Others 4. High school and equivalent 8. Doctorate and equivalent
P502	Language used at home and in the settlement (write down):
P503	Language used in formal institutions (school, workplace, write down):
P504	Community service in the last year (number); ....
P505	Neighbourhood watch in the last year (number); ....
P506	Folk/traditional party in the last year (number); ....
P507	Helping residents who have experienced a family death in the last year (number); ....
P508	Helping residents who are sick in the last year (number); ....
P509	Helping residents who had an accident in the last year (number); ....

Bibliography

Bappenas. no year-a. *15 Terrestrial Ecosystems Protecting, Restoring and Enhancing Sustainable Use of Terrestrial Ecosystems, Managing Forests Sustainably, Stopping Desertification, Recovering Land Degradation, and Stopping Loss of Biodiversity* [internet]. Available at: http://sdgs.bappenas.go.id/tujuan-15/.

Bappenas. no year-b. *14 Ocean Ecosystems Preserve and Sustainably Utilise Marine and Ocean Resources for Sustainable Development* [internet]. Available at: http://sdgs.bappenas.go.id/goal-14/.

BPS. 2020b. *Human Development Index (HDI) 2020.* [internet]. Available at: file:///C:/Users/BANGSIS-1/Downloads/BRSbrsInd-20201215113542.pdf.

BPS. 2020c. *Anti-Corruption Behaviour Index 2020* [internet]. Available at: https://www.bps.go.id/publication/download.html?nrbvfeve=ZTA3MjBkMDBjNmMwZjEzMGFmMWRjYTYx&xzmn=aHR0cHM6Ly93d3cuYnBzLmdvLmlkL3B1YmxpY2F0aW9uLzIwMjAvMDgvMzEvZTA3MjBkMDBjNmMwZjEzMGFmMWRjYTYxL2luZGVrcy1wZXJpbGFrdS1hbnRpLWtvcnVwc2ktMjAyMC5odG1s&twoadfnoarfeauf=MjAyMS0xMC0yNSAwNjoxNDowNw%3D%3D.

BPS. 2020d. *Percentage of Households by Province, Type of Area and Adequate Sanitation (Percent), 2018–2020.* [internet]. Available at:bps.go.id/indicator/29/834/1/percentage-of-households-by-province-type-of-area-and-adequate-sanitation.html.

BPS. 2021a. *Poverty and Inequality.* [internet]. Available at: https://www.bps.go.id/subject/23/poverty-and-inequality.html.

BPS. 2021b. *Percentage of Poor Population (P0) by Province and Area.* [internet]. Available at: https://www.bps.go.id/indicator/indicator/view_data/0000/data/192/sdgs_10/1.

BPS. 2021c. *Percentage of Poor Population September 2020 increased to 10.19 percent.* [internet]. Available at: https://www.bps.go.id/pressrelease/2021/02/15/1851/persentase-penduduk-miskin-september-2020-naik-menjadi-10-19-persen.html#:~:text=Jumlah%20penduduk%20miskin%20pada%20September,88%20persen%20pada%20September%202020.

BPS. 2021d. Poverty Profile in Indonesia, *in* Official Statistics News Vol 24 No 16/02.

BPS. N.d. Renewable Energy Mix. [internet]. Available at: https://sirusa.bps.go.id/sirusa/index.php/indikator/1563.

GIDRM. N.d. Local *Flood Early Warning System in the Philippines* [internet]. Available at: https://www.gidrm.net/en/products/local-flood-early-warning-systems

https://www.un.org/sustainabledevelopment/sustainable-consumption-production/

Huffpost. N.d. *What Do We Mean by a Quality Education?* [internet]. Available at: https://www.huffpost.com/entry/what-do-we-mean-by-a-qual_b_9284130

Institute for Essential Services Reform. 2019. *Sustainable Energy Access for Rural Communities: Status, Challenges, and Opportunities.* [internet]. Available at: http://iesr.or.id/wp-content/uploads/2019/05/Proceeding-PE-11.pdf

A. H. Iskandar, *The Methods of Measuring the National Sustainable Development Goals in Indonesia*, Engaging Indonesia, https://doi.org/10.1007/978-981-96-0927-7

KBBI. N.d. *Prosperous*. [internet]. Available at: https://kbbi.web.id/sejahtera

KemenPUPR. N.d.-a. *Potable Water*. [internet]. Available at: https://data.pu.go.id/dataset/air-minum-layak

KemenPUPR. N.d.-b. *Adequate Sanitation*. [internet]. Available at: https://data.pu.go.id/dataset/sanitasi-layak

Ministry of Health. 2020. *Regulation of the Minister of Health of the Republic Indonesia Number 2 of 2020 Child Anthropometry Standards* . Jakarta (ID): RI.

Minister of Maritime Affairs and Fisheries. 2017. *Decision of the Minister of Maritime and Fisheries of the Republic of Indonesia Number 50/KEPMEN-KP/2017 about Estimation of Potential, Allowed Catch Amount, and Level Utilization of Fish Resources in the Fisheries Management Area of the State Republic of Indonesia* . Jakarta (ID): RI.

UN Environment Programme. No year. Sustainable Consumption and Production Policies. [internet]. Available at: https://www.unep.org/explore-topics/resource-efficiency/what-we-do/sustainable-consumption-and-production-policies

UN. undated-a. *Goal 12: Ensure sustainable consumption and production patterns* . [internet]. Available at:

UN. undated-b. *Goal 15*. [internet]. Available at: https://sdgs.un.org/goals/goal15

UN. undated-c. *Goal 16: Promote just, peaceful and inclusive societies* [internet]. Available at: https://www.un.org/sustainabledevelopment/peace-justice/

UN. undated-d. *Life Below Water: Why It Matters*. [internet]. Available at: https://www.un.org/sustainabledevelopment/wp-content/uploads/2019/07/14_Why-It-Matters-2020.pdf

VVOB. undated. *VVOB's Definition of Quality Education* [internet]. Available at: https://www.vvob.org/en/education/our-vision-on-quality-education

WHO. undated. *WHO remains firmly committed to the principles set out in the preamble to the Constitution*. [internet]. Available at: https://www.who.int/about/governance/constitution

Weber, M. 1978. *Economy and Society*. California: University of California Press.

World Bank. 2009. *Handbook on Poverty and Inequality*, Washington DC, US: The International Bank for Reconstruction and Development/The World Bank.

Index

A. H. Iskandar, *The Methods of Measuring the National Sustainable Development Goals in Indonesia*, Engaging Indonesia, https://doi.org/10.1007/978-981-96-0927-7

MIX
Papier aus verantwortungsvollen Quellen
Paper from responsible sources
FSC® C105338

If you have any concerns about our products,
you can contact us on
ProductSafety@springernature.com

In case Publisher is established outside the EU,
the EU authorized representative is:
Springer Nature Customer Service Center GmbH
Europaplatz 3, 69115 Heidelberg, Germany

Printed by Libri Plureos GmbH
in Hamburg, Germany